Bone Tumors in Domestic Animals

Leonardo Leonardi

Bone Tumors in Domestic Animals

Comparative Clinical Pathology

Springer

Leonardo Leonardi
Department of Veterinary Medicine
University of Perugia
Perugia, Italy

ISBN 978-3-030-90209-4 ISBN 978-3-030-90210-0 (eBook)
https://doi.org/10.1007/978-3-030-90210-0

This Springer imprint is published by the registered company Springer Nature Switzerland AG.
The registered company address is: Gewerbestrasse 11, 6330 Cham, Switzerland

Prof. Alessandro Ciorba (10.09.1947–14.03.2021)

In memory of you Alessandro, that suddenly and prematurely left us. You left an indelible mark on my life because thanks to you I have learned to love pathology. You have been a mentor, a friend, and a gentleman of yesteryear. You did not give me time to share this work with you, but I am sure that you will still watch over my path, just as you did in your earthly life. Picture taken by L. Leonardi of his friend Alessandro Ciorba at the Department of Veterinary Medicine—University of Perugia—Italy.

Foreword to the First Edition

The author of this textbook has an infectious personality, enjoys teaching veterinary students and desires to share his extensive knowledge as a diagnostic pathologist with other pathologists and colleagues. This impression became immediately evident to me during the first of my invitations to present guest lectures on musculoskeletal pathology at veterinary colleges in Italy while I was a professor of pathology at the School of Veterinary Medicine, University of California at Davis. Later, he and his family paid a reciprocal visit to my wife and I while he and shared our experiences regarding our interpretation of bone tumors and orthopedic diseases of animals from our two countries. He brought veterinary students on short visits to my college on other occasions in which the students expressed their appreciation for his love of teaching.

Recently, when he inquired about illustrations for some uncommon bone tumors, I learned that he was drafting a textbook. However, that progress has been interrupted by the Covid-19 pandemic that was more severe in Italy than here in the USA. Soon he sent me an advanced copy of the draft and asked for my thoughts. After perusing this work, it was apparent that this textbook reflected his love for teaching veterinary students while he providing them with a basic understanding bone tumor biology and a description of the musculoskeletal tumors. For the experienced investigator the textbook also provided appropriate special techniques that they may be required for a precise diagnosis.

For veterinary students and residents, his textbook provides basic information about origin, development, and physiology of cells, cellular matrices and formation of structural elements of the skeleton prior to presenting the fundamental diagnostic features of musculoskeletal tumors and tumor-like lesions. In addition to providing diagnostic features of each entity, he often includes historical information of earlier investigations that led to tumor origin and diagnosis not found in other veterinary textbooks dealing with bone tumors. His textbook offers comparisons between musculoskeletal tumors in man and animals in the respective World Health Organization publications. The book contains charts with differential diagnoses of similar tumors including information on special staining techniques. He emphasizes the use of imaging technologies in reconciling those findings with microscopic findings from tissue specimens to be certain if the surgical specimens are representative samples from the most diagnostic sites. Therefore, this textbook is not only a good source for the student but a quick reference for the experienced pathologist and clinician.

School of Veterinary Medicine,
University of California, Davis, CA,
USA

College of Veterinary Medicine and
Biomedical Sciences, Texas A&M
University, College Station, TX, USA

Roy Pool

Preface

> I don't want to feel intelligent by looking at stupid people, I want to feel stupid by looking at intelligent people
>
> (Franco Battiato 1945–2021)

During my professional career, I had the opportunity to meet colleagues and masters who for me have represented sources from which I've learned science.

Together with them, I managed to mature on an often tortuous and difficult, sometimes almost impossible, path that allowed me to gain experience and gave me courage in tackling the creation of this book for those who want to dedicate themselves to bone pathology.

Allow me to recall these points of light that have illuminated my path as a man first and then as a pathologist.

It is not easy to be able to express certain feelings, and I'm not even sure I can, but the bases of human and social relationships represent for me the essential and indispensable elements of every form of human sharing, even the scientific ones. I have based my whole life on this; I would like to continue on this path scientifically as well.

The first thank you is for the family from which I come and where I learned the human and social foundations of respect and attention to others. To my beloved father Aldo and my beloved mother Elvira, my brother Luca, my deep, grateful, and pure love.

The tree and the sweet fruits that I have been able to cultivate and grow in my life with the creation of my family, my only and pure reason for living, are firmly linked to my roots. In Monica, Francesco, and Alberto, I have projected every effort to improve my being, dedicating to them every moment of life we have lived together. They represent energy, life, depth, example, love.

The mediocre teacher tells. The good teacher explains. The excellent teacher demonstrates. The master inspires (Socrates). Socrates, on the other hand, helps me to mention with a sense of deep gratitude for those masters of Science and life who have helped and inspired me in the long journey of my cultural and continous scientific growth. Starting from the earliest years, the first thanks is for Prof. Alessandro

Ciorba, who gave me the opportunity to start learning about histopathology and veterinary sciences. It is his "fault" that I became an academician and a pathologist, with all the consequences that this choice may have determined. I have rarely met professors of his elegance and intelligence. Thanks to my teacher, Prof. Italo Manocchio, who always trusted me, making me believe that I too could give my humble contribution to the scientific world.

Thanks Prof. Giovanni Di Guardo. We shared life and science experiences with you like no one else I will ever be able to do together. We are united by feelings of mutual esteem and respect that also involve our families and which we will no longer be able to do without.

Thanks to Prof. Franco Roperto who honored me with his friendship and allowed me to collaborate in his important research, teaching me the salient aspects of true international research and allowing me to be co-author of many of his works.

A heartfelt thanks to my dear friends Dr Deborah Gillette and Dr. Corrie Brown (UGA), my American "sisters" who helped me to open the way to internationality, helping me to find the way to America, from which I continued my wandering to many other parts of the world. To they I really owe a lot in terms of gratitude and above all of deep affection and love.

Last but certainly not least, one of the most important people who, especially in this scientific context, rises to the exclusive and sole point of reference, Prof. Roy R. Pool, Emeritus Professor UCD and Texas A&M (USA). In addition to being a close friend, he is also the one I consider to be the number one and undisputed international authority in veterinary musculoskeletal pathology. Roy was a point of inspiration and reference for me even before I met him. My experience with him at Texas A&M University represented a starting point for a systematic and specialized study of bone lesions in pets. Roy is a special person, his kindness and his availability led me to emulate his approach to life, which still represents a source of daily inspiration. Roy has supported me since 2004 when, thanks to sister Debbie, I had the opportunity to meet him and plan a period of study and knowledge with him. Since then, we have never lost sight of each other, always staying in touch over the years, together with his wife Bettye. Thank you, Roy, for your support, for your teachings, for your friendship. His suggestions, his corrections, his lessons were fundamental for the elaboration of this text.

Thanks also to Springer Nature Group to give me this important opportunity.

And "last impersonal" thank you, more generic, but equally profound, is for the University of Perugia, my University, and for my students. In the University, I grew up as a man and as a small researcher, from it and in it I received unique opportunities engraved indelibly and deeply in my heart and mind. A deep thank you to my beloved Students, both of Veterinary Medicine and Biotechnology, for always encouraging me to improve. I hope to have been a positive example in their approach to their studies and the professional world of work.

To all those who will be able to use it, I leave this book in the hope that it will be useful in dealing with the complex and dynamic world of animal bone pathology.

Perugia, Italy Leonardo Leonardi

Acknowledgement

"Me and Roy," I would like to pay a tribute to Dr. Roy R. Pool, Emeritus Professor at UCD and Texas A&M, my mentor, my exceptional friend, and my great supporter for this book.

Dr. Roy R. Pool is a giant of veterinary bone pathology, a unique example of how to do science and life, a humble man, kind and good, a scientist and an inexhaustible well of knowledge from which I have gathered inspiration and without whom I would never have written this book. Roy and my family, Monica, Francesco, Alberto and also Roy's wife, dear Bettye Pool, have been my real engine and my strength to realize this text. We are all together, as only one family.

Contents

Chapter 1
Outline of Anatomy/Physiology of Skeletal Tissues

1.1 Structure and Physiology of Skeletal Tissue

1.1.1 Bone Tissue

Bone is a living and active tissue in a constant phase of modeling and remodeling and, together with cartilage, it is the main component of the tissues of mesenchymal origin responsible for supporting the body (Bani 2007; Adler 2000; Hall 2005). Bone and cartilage are fundamental components of the skeleton that carry out important and multiple functions, including allowing mobility, protecting many internal organs, containing the bone marrow, and an important calcium reserve which, thanks to the continuous activity of deposition and of mobilization regulated by sophisticated endocrine mechanisms, guarantee the regulation of the calcium ion in the plasma. The origins of bone tissue date back to more than 300,000 years ago and its main feature is represented by the fact that it is mineralized, and it is composed of a significant number of intercellular substances very rich in calcium phosphate. In vertebrates, four different types of mineralized tissue are identified which are represented by bone, cartilage, dentin, and enamel, although the latter two are dental tissues (Hall 2005).

The skeleton is essentially composed of two types of tissues represented by bone and cartilage and by four types of cells: preosteoblasts or progenitor cells, osteoblasts, chondrocytes, and osteoclasts. Bone densit types are spongy and compact bone (Farnum and Wilsman 2002; Farnum et al. 2002, 2003). The short bones, the flat bones, and the epiphyseal structures of the long bones contain an abundance of spongy bone which is named for its structural porosity resembling that of sponge, rich in bony trabeculae that are oriented to resist mechanical forces. The cancellous bone borders the medullary cavity and because its porosity contains bone marrow where it performs an important hematopoietic function. The superficial or cortical portions of short bones, long bones, and flat bones are made up of compact bone. Compact bone is made up of calcified material referred to as mineralized

L. Leonardi, *Bone Tumors in Domestic Animals*,
https://doi.org/10.1007/978-3-030-90210-0_1

extracellular matrix that is arranged in layers, i.e. lamellar bone or in higher animals that also have longitudinally oriented cylindrical bony units called osteons. Osteocytes are buried in the calcified matrix of compact bone. The cylindrical osteons contain a central cavity where the bone lining cells and buried osteocytes are nourished by a longitudinally oriented blood vessel. The central cavity is called a Haversian canal and the central blood supply is the Haversian vessel.

Continuous bone remodeling activity throughout life of the vertebrate permits removal of bone damaged by use and its replacement by viable bone tissue. Remodeling with removal of effete bone and its replacement by new viable bone allows the bony structure to modify its shape to adapt to changing mechanical forces required for support, locomotion, protection, injury, or other pathologic insults and for changing nutritional and physiological stimuli.

Rate of bone deposition determines the two histologic pattern of bone tissue. Bone tissue deposited at more than a 1 mm per day is woven bone. It has randomly deposited collagen fibers that entrap osteocytes whose cell processes are not connected to osteoblasts on the newly forming surfaces of the woven bone. By comparison lamellar bone is formed more slowly allowing the collagen fibers to be deposited in an orderly alignment related to the forming bone surface. As osteoblasts form and become entrapped in the forming bone surface and become osteocytes, their thin elongated cytoplasmic processes maintain connections with those of the newly deposited osteoblasts that cover the newly forming bone surface of lamellar bone. In this manner the cellular array of buried and surface osteogenic cells maintain connections in a canalicular–lacunar system and participate in calcium homeostasis of the individual as well as structural support.

Bone is, therefore, a vascularized skeletal tissue that can also arise ectopically outside the skeleton in other parts of the body (Karsenty 2003; Kronenberg 2003; Schuenke et al. 2007).

1.1.2 Bone Cells

Admixed with the extracellular matrix the four main cellular entities are represented by preosteoblasts or osteoprogenitor cells, osteoblasts, osteocytes, and osteoclasts (Takayanagi et al. 2002). Osteoprogenitor cells, or preosteoblasts, are oval-shaped and/or polygonal-shaped cells with a scarce weakly basophilic cytoplasm and nuclei containing large, prominent nucleoli. These cells are mainly located in the layer of periosteum on the cortical surface or in the endosteal lining of medullary surface of cortex bone. Osteoprogenitor cells produce and secrete autocrine growth and differentiation factors, such as Bone Morphogenetic Protein (BMP) which participate in the activation of cell differentiation phases that lead to the formation of mature osteoblasts.

Osteoblasts are cells that originate from stromal cells of the bone marrow and which perform the function of producing specific proteins that contribute to forming the so-called bone matrix (also called osteoid) and its mineralization. Osteoblastic activity is strongly conditioned by various stimuli such as that of parathyroid hormone (PTH) and vitamin D and vitamin C. When osteoblasts proliferation terminates, then initiate production of extracellular matrix, and become osteocytes. Inactive osteoblasts flatten into bone lining cells that cover bone surfaces. The bone lining cells maintain connections with other adjacent bone lining cells by intercellular cytoplasmic junctions. The lining cells also extend thin elongated cell process that contact at gap junction sites with cytoplasmic processes of underlying recently buried osteocytes that travel through canaliculi in the bone mineralized bone matrix. This interconnecting pathway permits the bone lining cells and buried osteocytes to participate in metabolic functions in a lacunocanalicular network called the canalicular–lacunar system.

Osteoblastic cells have large dimensions (about 20 μm) and a polyhedral shape and a tendency to connect to each other to form laminar structures close to the surface of the bones. The nuclei of osteoblastic cells are roundish in shape, euchromatic, with a large significantly basophilic cytoplasm. In the osteoblastic cell the nucleoli generally appear large and well evident and there may also be cytoplasmic granules better detectable with PAS staining and Schiff's reagent and which represent the morphological expression of secretory vesicles and intracellular calcifying globules. Osteoblasts are also usually positive for alkaline phosphatase reaction. Osteoblastic cells also tend to connect with other osteoblasts and with osteocytes by means of gap junctions which also serve to interchange signals for the control of bone metabolic activities. The various metabolic phases of bone certainly undergo a strong control also and above all by the osteoblastic cells that produce the bone matrix which mineralizes according to well-determined and sequential rules of arrangement and orientation.

Among the soluble factors secreted by osteoblasts there are the transforming growth factor ß (TGF- ß), the insulin-like growth factors (IGF), and the enzymatic and osteoclast activating factors such as collagenase and tissue plasminogen activator (tPA) (McSheehy and Chambers 1986; Gorski et al. 1990; Tezuka et al. 2002; Nobta et al. 2005).

Osteocytes are terminal and representative cells of mature bone tissue.

Osteocytes are located inside the osteocyte lacunae and are influenced by exocrine and endocrine stimuli such as PTH, vitamin D, and Vitamin C. They are stellate cells with numerous thin cytoplasmic extensions that travel through the canalicular system to connect to bone lining cells

Osteocytes are characterized by a poor basophilic cytoplasm, heterochromatic nuclei, and small nucleoli.

Osteoclasts are multinucleated cells of myeloid lineage whose function is to participate in shaping bone during skeletal development, bone remodeling, and bone repair of the skeleton by eroding bone surfaces. Osteoclast derives from preosteoclasts, that is, from monocyte cells of the blood and hematopoietic bone marrow that have macrophage activity and that have a specialized lining, a ruffled border, that binds matrix adhesion proteins for resorbing bone surfaces, feature not present in macrophages. Osteoclasts are large (100–200 μm) and can contain up to

50 nuclei, with little chromatin and evident single nucleoli with cytoplasm weakly eosinophilic.

They are responsible for the destruction of the bone through an enzymatic digestion process that involves the formation of resorption pits, called Howship's lacunae, and are sites of resorbed bone surfaces which take approximately 3–7 days to complete. After this time and a short additional transition period the resorption pits are filled with osteoblastic cells that begin to apply bone osteoid matrix over the resorption pit that progressively mineralizes and forms a bone surface.

This process takes about 10–15 days and the osteoclastic activity is conditioned by various factors which include PTH, vitamin D, calcitonin, the RANK-RANKL-OPG system, PG2, etc. (Akatsu et al. 1989; Suda et al. 1992, 1995; Quinn et al. 1994).

A peculiarity of osteoclasts, likely because of their hematopoietic origin, they are the only cells among bone tissue associated cells that has receptors for calcitonin (inhibitor of bone resorption), a hormone produced by the parafollicular cells (or C cells), of the thyroid and which has a strong antagonistic power to the parathormone. Osteoclasts do not have the receptor for parathyroid hormone.

There is a close physiological correlation between osteoblasts and osteoclasts also linked to the activation of soluble cellular factors called osteoclast activating factors (OAF) that participate in the resorption of bone tissue, together with other factors such as the granulocyte-macrophage colony stimulating factor GM-CSF.

Other cellular elements involved in the stimulating dynamics of bone cells are represented by endothelial cells which, by producing soluble factors such as IGF (insulin-like growth factor), stimulate osteoprogenitor cells to differentiate into osteoblasts.

These same endothelial cells also activate chemotactic factors for production of osteoclasts.

Other factors that can affect bone cells are represented by various interleukins (Il-1, IL-3, IL-6), tumor necrosis factor (TNF), platelet-derived growth factor (PDGF), epidermal growth factor (EGF), etc. (Horton et al. 1984; Baron et al. 1986; Udagawa et al. 1990; Shinar and Rodan 1990).

1.1.3 Bone Tissue Organization

Bone tissue is a connective tissue that contains a large amount of extracellular organic matrix relative to its cellular component. It is a composite of collagenous and other proteins embedded in a mineralized intercellular material of calcium, phosphate, and carbonate. The bone's fibers present in the bone are almost predominantly made up of collagen fibers (type I collagen). Sharpey's fibers which also contain small amounts of elastic fibers firmly adhere the periosteum to the bone surface. The intercellular bone substance contains variable amounts of type I (PG-I), type II (PG-II) proteoglycans and glycoproteins, the most abundant of which is represented by the osteonectin responsible for several calcium concentration processes. Osteonectin represents another important adhesion glycoprotein capable of binding to collagen and connecting it to bone hydroxyapatite, also conditioning physiological and pathological processes of organization of bone cells and their migration and adhesion to the osteoid matrix.

Another bone glycoprotein is alkaline phosphatase, an enzyme also of significant diagnostic importance in oncology, whose normal function seems to be that of involvement in mineralization processes or, according to only some authors, in the synthesis of bone matrix.

Among the various proteins with peculiar and fundamental characteristics for the physiological dynamics of bone, we must emphasize the importance of osteocalcin also called bone GLA protein (containing γ-carboxyglutamic acid). Osteocalcin seems to have an important regulatory function towards the calcium to which it binds, thereby preventing its availability to bind with phosphate and, therefore, to form bone mineralization. It, therefore, represents an inhibitor of bone mineralization.

As reported in humans by S.H. Harada & G.A. Rodan in Nature: "*the skeleton is an efficient "servo" (feedback-controlled/steady-state) system that continuously integrated signals and responses which sustain its functions or delivering calcium maintaining strength. In many individuals, bone mass homeostasis starts failing in midlife, leading to bone loss, osteoporosis and debilitating fractures* . . . ".

There are different types of bone development that always grows due to a form of replacement of an existing tissue, through a process that takes the name of ossification or osteogenesis. Basically, there are two types of processes of bone formation by osteoblasts that are responsible for ossification of the skeleton: (1) intramembranous ossification, a process in which bone tissue applies by the periosteal membrane to a bone surface, e.g. flat bones of the skull or surfaces of other bones of the skeleton formed by endochondral ossification and (2) endochondral ossification, a process in which bone tissue develops from a cartilage model of most bones of the appendicular and axial skeleton except for the skull. Synonyms sometimes used for intramembranous bone are direct or appositional bone while synonyms for endochondral ossification are indirect or chondral ossification (Adler 2000; Hall 2005).

1.1.4 Normal Bone Development

Bones grow and shape and remodel, especially during the growth phase, even it is now known that bone undergoes modifications and reworking throughout the life of an animal.

Bone growth basically occurs in length and width and is significantly influenced in its formation by various endocrine and metabolic factors.

Long bones grow in length by endochondral ossification (Fig. 1.1) in which the rate of interstitial growth of cartilage in the deep layer of the articular cartilage that forms the epiphyseal cancellous bone and in the growth plates at the ends of these bone is replaced at that same rate by metaphyseal cancellous bone in a normal

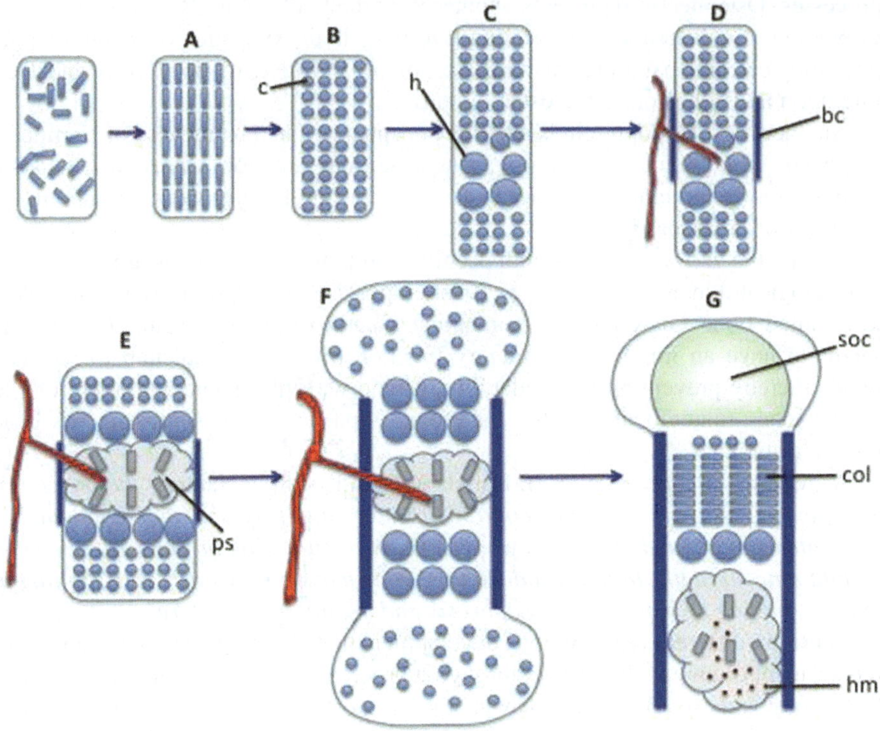

Fig. 1.1 Endochondral bone formation. (**a**) Mesenchymal cells condense. (**b**) Cells of condensations become chondrocytes (c). (**c**) Chondrocytes at the center of condensation stop proliferating and become hypertrophic (**h**). (**d**) Perichondrial cells adjacent to hypertrophic chondrocytes become osteoblasts, forming bone collar (bc). Hypertrophic chondrocytes direct the formation of mineralized matrix, attract blood vessels, and undergo apoptosis. (**e**) Osteoblasts of primary spongiosa accompany vascular invasion, forming the primary spongiosa (ps). (**f**) Chondrocytes continue to proliferate, lengthening the bone. Osteoblasts of primary spongiosa are precursors of eventual trabecular bone; osteoblasts of bone collar become cortical bone. (**g**) At the end of the bone, the secondary ossification center (soc) forms through cycles of chondrocyte hypertrophy, vascular invasion, and osteoblast activity. The growth plate below the secondary center of ossification forms orderly columns of proliferating chondrocytes (col). Hematopoietic marrow (hm) expands in marrow space along with stromal cells. (Adapted from Henry M. Kronenberg, Nature 2003)

animal. However, growth in width of the bone results from deposition of circumferential layers of lamellar bone by intramembranous ossification deposited by the periosteum.

The increase of the bones in width is essentially due to the apposition of bone from the periosteal region, after bone neovascularization and activation of mesenchymal-type cells that differentiate into osteoblasts, forming primitive and perivasal osteonal fibrous sketches which, subsequently, will be replaced with bone of lamellar type. Once the bone has reached its final size, endosteal osteoblastic cells form the bone lamellae and a thin layer of trabecular bone (Kronenberg 2003).

All these growth phenomena summarized briefly are significantly influenced by the action of numerous factors such as:

- Growth hormone or somatotropin (GH) produced at the pituitary level, stimulates the production of somatomedins (proteins produced mostly in the liver promoting cells growth) with, growth and prometabolic factors in the liver against chondrocytes and proliferating cartilage. Inadequate formation of this hormone is responsible for diseases such as "pituitary dwarfism" in humans and which in animals, especially dogs, recognizes an autosomal hereditary disease that affects various breeds such as German Shepherd, Dachshund, Basset Hound, etc. An excess of GH production during the growth phases can lead to diametrically opposite diseases, such as that of gigantism.
- Parathyroid hormone (PTH) produced by the parathyroid glands, acts by determining the proliferation and differentiation of osteoblasts and, more sporadically, the activation of osteoclasts by stimulation of the osteoclast activating factor (OAF). PTH, therefore, promotes the raising of calcium levels and the reabsorption of the same in the kidney: hypercalcemizing effect.
- Calcitonin produced by the parafollicular cells, (C cells) located in the thyroid gland helps regulate levels of calcium and phosphate in the blood and opposes the action of parathyroid hormone by reducing levels of calcium in the blood while inhibiting bone resorption by osteoclasts.
- Vitamin A, fat-soluble with directly inhibiting action on osteoblastic production and increasing the expression of vitamin D receptors.
- Vitamin C is water-soluble and is strongly involved in the synthesis of collagen and is also a cofactor for osteoblasts, especially in those responsible for the synthesis of collagen in the bone matrix.
- Vitamin D. It is mainly taken with the diet. It promotes osteoblastic differentiation, stimulates the production of bone matrix and the deposition of bone calcium. In the intestine, it stimulates the absorption of calcium, while inhibiting its elimination in the kidney. Its deficiency is responsible for mineralization deficits that lead to rickets in young subjects or osteomalacia in adults' patients.
- Sex hormones. Estrogen and testosterone stimulate differentiation and functionality of osteoblasts and also have an inhibitory action against the chondrocyte growth of proliferating cartilage, which causes the epiphyses to close, thus also blocking bone growth.
- Thyroid hormones (T3 and T4) promote cell metabolism and bone maturation.
- Nitric oxide (NO) is produced by many cells, especially endothelial ones, and is able to stimulate osteoblastic differentiation, also attributing a potential bone formation by stem cells located in the endothelial lining of arteries.
- Molecular oxygen is an important element for oxidative phosphorylation and for the stimulation of osteogenesis.

Calcium plays a fundamental role in many physiological and cellular activities of the organism (muscle, bone, nervous system, etc.), in the metabolism of some hormones, in coagulation factors, in the activation of enzymes, in the control and dynamics of bone and tooth metabolism. The greatest amount of calcium is certainly

Table 1.1 The main non-collagen proteins of the bone matrix (Adapted from Hall B.K., 2005, Elsevier)

• OSTEOCALCIN: Osteocalcin (γ-carboxyglutamic acid, Gla, bone protein Gla, BGP) is a protein, Ca ++ binding, vitamin K-dependent, of 5800 MW, containing γ-carboxyglutamic acid. Among the seven most abundant proteins in human bone, osteocalcin is the most represented type of non-collagen protein with a percentage of 10–20% of non-collagen proteins. Its levels in human serum are 7.0 ± 2.5 ng/ml. Osteocalcin activates osteoclasts and precursors of osteoclasts in the bone for its resorption.
• OSTEONECTIN: Osteonectin (SPARC, secreting protein, acidic, rich in cysteine) is a 32,000 MW protein of the extracellular matrix that represents 10% of bone proteins. It appears related to the mineralization process; Osteonectin binds collagen to hydroxyapatite, serves as a mineralization core, and regulates the formation and growth of hydroxyapatite crystals.
• OSTEOPONTIN. Osteopontin is a 66-kDa glycosylated phosphoprotein synthesized by osteoblasts, osteoclasts, and macrophages. It is easily found in sites where bone metabolism is activated: endochondral bone and membrane bone, osteoid, preosteoblasts, osteoblasts, and osteocytes. Osteopontin increases the cell survival index and the cell migration index but decreases by inhibiting the mineralization process.

contained in the bone tissue which is in the body, it is also found in small and variable quantities in intracellular and extracellular sites.

Calcium homeostasis is, therefore, controlled at various levels which, in addition to bone tissue, also include other organs and systems such as the intestine and kidneys. Complex regulation of calcium involves various endocrine and exocrine factors that include parathyroid hormone secreted by the parathyroid glands, calcitonin secreted by parafollicular (C-cells) present in the thyroid gland, and vitamin D by the skin and liver interaction.

Another important factor that regulates the structural and functional activities of cell membranes and organelles is phosphate. Phosphate is also a fundamental element in the composition of both bone and teeth, but phosphate is also involved in the regulation of other biological phenomena such as pH regulation. Phosphate is also regulated according to control pathways similar to those of calcium. The mineral component of bone consists of calcium salt crystals mainly represented by calcium phosphate and, to a lesser extent, by calcium carbonate, magnesium phosphate, and calcium fluoride.

Calcium phosphate also represents the component called "apatite crystals" which by binding with hydroxyl ions forms hydroxyapatite, a fundamental component of bone tissue (Table 1.1).

1.1.5 Types of Skeletal Tissues

Bone tissue is a mineralized tissue whose origin seems to date back to more than 250,000–300,000 years ago. In vertebrates, the types of mineralized tissues are essentially represented by four different types: bone, cartilage, dentine, and enamel. Dentine and enamel are said to have originated in early vertebrates from endoderm that gave rise to pharyngeal teeth of jawless vertebrates. With evolution of the jaws in these primitive vertebrates the pharyngeal teeth moved forward with the evolution of the jaws. Others indicate that their resemblance to shark teeth indicate that etiology suggesting they are of extraskeletal origin. Whatever, the case vertebrae teeth did not have the same embryological origin as did mesodermal derived bones of the vertebrate skeleton. Therefore, dentine and enamel are regarded as "dental tissues." Bone arose as a primitive tissue from mesoderm that gave rise to muscles, attachments, and an endoskeleton. The endoskeleton provides locomotion and protection of internal organs in comparison to the exoskeleton of the shells of crabs and insects that also allows movement through attachment of muscles to the exoskeleton while providing protection and support of the body.

In invertebrates the exoskeleton or the cuticle is mainly composed of chitin (polysaccharide consisting of different units of N-acetylglucosamine, very similar to cellulose), often mixed with glycoproteins, sometimes mineralized, or containing parts of calcium carbonate, but rarely characterized by presence of calcium phosphate, which, on the other hand, is the main component and most present in bone structures. Even non-mineralized bone cartilage is not present in many invertebrates, despite some of their chondroid components showing similarities and biomolecular characteristics similar to those of bone. By suggesting to the reader, the deepening of the study of specialized anatomy and bone biology texts, we would like to remember that in invertebrates the mineralization process of cartilage seems to be strongly conditioned and blocked above all by temperature, also associated with other factors such as low calcium concentration and of phosphorus particularly frequent especially in animals living in the marine environment.

The mineralization process is not even an exclusive property of the four mineralized tissues of vertebrates and is also a detectable process in metazoans and some single-celled organisms. The table below extrapolated from the text by B.K. Hall reports some diversity of different mineralized biological tissues in some groups of organisms and the main organic components associated with mineralization (Table 1.2).

1.1.6 Bone and Cartilage Skeletal Tissues

1.1.6.1 Bone

Bone is a vascularized tissue that constitutes the supporting structure of the vertebrate body, that is, the skeleton. It is made up of different types of cells and a mineralized extracellular matrix (ECM). Bone cells are mainly represented by osteoblasts and osteocytes and osteoclasts. Type 1 collagen, composed of two αl chains and one αll chain (products of the respective genes Col1a1 and Col2a1) and described as α1 (ll) 3, represents the main component of the extracellular matrix.

Table 1.2 Definitions and terminology for skeletal systems and ossification models (from Hall, B. K., 2005)

Structure	Definition	Model of Ossification
EXOSKELETON	Skeletal system that forms in contact with ectoderm and endoderm; dermal bone, scales, fins, gills, teeth.	INTRAMEMBRANOUS
ENDOSKELETON	Skeletal system that does not form in contact with the ectoderm and the endoderm; large chondral bone.	ALL TYPES
CHONDRAL BONE	Type of bone that develops into a pre-formed cartilage model; the major components of the endoskeleton.	ENDOCHONDRAL
PERICHONDRAL BONE	Subtype of chondral bone that ossifies from a perichondral connective tissue.	PERICHONDRAL
MEMBRANOUS BONE	Type of bone that mesodermal mesenchyme	INTRAMEMBRANOUS
SUBDERMAL BONE	Type of newly formed bone, which develops without contact with the ectoderm or endoderm; sesamoid bones, pathological ossifications.	INTRAMEMBRANOUS
PERICORDIAL BONE	A type of bone that forms in the connective tissue bundles surrounding the notochord.	PERICHONDRAL
DERMAL BONE	Type of bone that is not pre-formed in the cartilage and that develops in contact with the ectoderm and the endoderm; the majority of the components of the exoskeleton.	INTRAMEMBRANOUS

Osteocalcin (bone protein gla or BGP), osteopontin, and osteonectin are, on the other hand, the most significantly expressed proteins in the bone. The bone matrix is permeated with numerous channels, also referred to as canaliculi, which represent intercellular connecting extensions that allow them to connect to other osteocytes, osteoblasts, and more superficial osteogenetic cells of the bone through junctions tending to form syncytia.

The first bone matrix produced and deposited is generally non-mineralized and is known by the name of osteoid. Subsequently, the osteoid merges mainly with hydroxyapatite to form the actual mineralized tissue. An organic lamina, defined as the lamina limitans, tends to separate the osteoid from the already mineralized bone. Bone is an aerobic tissue with a high oxygen consumption and is present only in vertebrates.

1.1.6.2 Cartilage

Cartilage is another important tissue closely related to bone. The chondroid tissue has the peculiarity of not being vascularized. As can also happen with bone tissue, cartilage can form in an ectopic form outside the skeleton, such as occurs in connective tissue, muscles, and heart. Cartilage originates from "chondroid" cells that identify with chondroblasts and chondrocytes and can be removed from both mononuclear and multinucleated cells called chondroclasts. Unlike bone cells, chondrocytes have no cellular connection processes and only some of them turn out to be ciliated.

The cellular matrix of vertebrate cartilage is mainly composed of glycosaminoglycans (GAGs), in which chondroitin sulfate and other different proteoglycans are particularly abundant. The largest amount of collagen present in cartilage is represented by type II collagen, which in turn is composed of three α II chains, described as α1 (I) 2 αII. Some types of vertebrate cartilage also contain additional collagen: type I in the joints, secondary cartilages and fibro-cartilages and type X in hypertrophic cartilages.

1.1.6.3 Dentine

Dentin is a mineralized dental tissue that forms the scaffolding of the teeth. Dentin is a primary tissue of the vertebrate exoskeleton and also an odontogenic and skeletal tissue. It consists of a prevalent percentage of inorganic material, mixed with smaller quantities of organic and liquid material. Dentin is made up of cells called odontoblasts that synthesize and lay a substance represented by an organic matrix called "predentine," which later mineralizes to actual dentin, in which there is also an evident amount of hydroxyapatite. Dentin is reabsorbed by odontoclasts and is not present in invertebrates.

Cartilage resists pressure, has a high fluid content, is anaerobic tissue and therefore has a low oxygen consumption. The functions of cartilage are to form the tissues of the embryonic endoskeleton of vertebrates (primary cartilage) and of the exoskeleton in some invertebrates.

Primary tissue of the vertebrate jaw of endodermal origin that in evolution arose from the pharyngeal teeth of jawless vertebrates that is of odontogenic tissue located in a socket of skeletal tissue of mesodermal origin. However, other reports indicate their similarity to shark teeth suggest that etiology.

1.1.6.4 Enamel

Enamel is the external avascular component of teeth and some scales, and is one of the highest levels of mineralization of the biological components of an organism with extremely hard characteristics. Enamel is made up of 96% inorganic matter and only 0.5% of organic matter. Enamel is produced by specific specialized cells called ameloblasts and in mammals its formation (enamelogenesis or amelogenesis) involves two phases: the deposition of the organic matrix and the mineralization of the matrix. Enamel is not found in invertebrates.

1.1.6.5 Intermediate Tissues

Intermediate tissues are those that may possess characteristics of two or more of the four listed skeletal tissues. Intermediate tissues are classified as cementum, enamel, chondroid, and chondroid bone. Some intermediate tissues, which we will not mention, are believed to be pathological tissues.

1.1.6.6 Cementum

Cementum is a tissue that contributes to the adhesion of the teeth of some vertebrates to their relative natural cavities. Cementum has the biochemical characteristics of dentin and calcified cartilage. It is deposited by cells called cementoblasts which, as for the odontoblasts that deposit dentin, the osteoblasts that deposit the alveolar support bone, the fibroblasts that lay the pulp and the periodontal ligament, originate from the dental papilla, which originates from the neural crest. The cementum is placed on the existing dentin.

1.1.6.7 Enameloid

Mineralized and aprismatic tissue, which lines the teeth and which is deposited following the combined action of ameloblasts and odontoblasts.

1.1.6.8 Chondroid and Chondroid Bone

These two components possess characteristics of cartilage and bone and resemble, to varying degrees, cartilage and bone in their development. Typically, both chondroid and chondroid bone have type I and II collagen in their matrix. Articular and secondary cartilages also have both types of collagens.

The chondroid bone matrix is basophilic as in bone, but has cell nests contained in lacunar structures that resemble chondrocytes. In mammals in particular, the chondroid bone is a permanent bone tissue. In this sense, the chondroid bone represents a stable intermediate tissue of the skeleton and not a transition tissue. Primary cartilage is a transitional tissue that replaces bone and is the least permanent part of the skeleton, compared to chondroid bone.

The chondroid is the minor permanent tissue. In the antlers of some deer (Odocoileus virginianus) the chondroid is replaced by bone and this process is called chondroid bone formation, an intermediate process between an intramembranous ossification and an intrachondral ossification. In other situations, the chondroid is remodeled and transformed into bone.

The division between bone and cartilage dates back historically to the time of Aristotle (384–322 BC). Already in the seventeenth century it was thought that cartilage could be transformed into bone. A student of the anatomist Gabriele Falloppio, Volcher Coiter, at the University of Padua was the first to fully describe the human skeleton, in 1559. Over the centuries, several scholars have argued, demonstrating it, that cartilage can turn into bone during the life of animals. It was around 1850 that W. S. Sharpey formulated his "Replacement Theory" in which he stated that bone replaces cartilage in endochondral bone.

Today, the difficulty of distinguishing cartilage from bone has posed the problem to three different medical and veterinary professional investigators, developmental biologists and zoologists.

Paleontologists have identified these tissues and determined their relationships through origin, evolution, and diversification in vertebrates. Pathologists often find that skeletal tissues cannot be definitively classified. Developmental biologists who evaluate cellular differentiation and skeletal tissue development are often frustrated when they compare and attempt to extrapolate their findings in long bones of non-mammals to long bones of mammals. This is caused by their encountering of a spectrum of intermediate matrices between cartilage and bone in non-mammals. From the years 1691–1693 with the invention of the optical microscope, rudimentary sections of bone tissue were examined by Havers and Leeuwenhoek, who essentially described bone microstructure. Both described tubules crossing and traversing a solid structure.

Early in the eighteenth century Duverney demonstrated that some of the "holes" previously demonstrated by Havers and Leeuwenhoek contained nerves and this was evidence that bone is not inert but an active structure. Clopton Havers, a physicist, identified these "holes" as tubules known today as Haversian canal. Much smaller channels were called canaliculi through which osteocytes sent cytoplasmic processes to communicate with one another.

It was Duverney who, in the early eighteenth century, showed that some of those "holes," previously identified by Havers and Leeuwenhoek, contained nerves as evidence that bone is not an inactive, inert tissue. Most of these "holes," on the other hand, represent the tubules, better known today as Havers canals, after the discovery of the physicist Clopton Havers, or the canaliculi in which the course of osteocytes takes place.

John Howship was the first to observe gaps in the surfaces of bone from histopathology bone specimens he examined. Today these focal semilunar sites or resorption bays caused by osteoclasts are called Howship's lacunae. Haversian system consists of a cylinder of bone tissue with a central Haversian cavity and a central Haversian vessel. An osteon is a term used to describe mechanical and metabolically functional cylindrical unit of bone. An osteon is composed of the Haversian system and bone cells in the wall of the Haversian system. The wall is formed of osteogenic cells called osteoblasts that line the surface of the canal and osteocytes embedded in osteocyte lacunae in the bony wall. Osteocytes communicate by thin cytoplasmic process that travel through a canalicular–lacunar system and connect to the bone lining cells. This system not only insures maintenance of osteocyte viability but their participation in metabolic functions regarding calcium homeostasis of the animal.

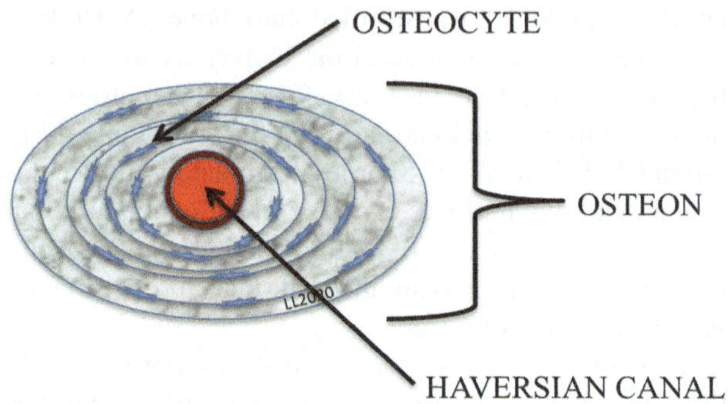

Fig. 1.2 The basic structural osteon system

They represent sites of resorbed bone in which discrete quantities of osteoclasts accumulate. Haversian systems (osteons) represent integrated units of bone structures (Fig. 1.2).

Over the years the preparation and observation techniques continued to improve, until 1846 when John Quekett standardized a new methodology which, thanks to the use of hydrochloric acid, allowed the mineral matrix to be removed from bone tissue samples. He practically standardized the demineralization or decalcification method that is still normally used today for the preparation of bone tissue to be examined microscopically.

At the same time, in the year 1845, John Goodsir demonstrated the presence of particular bone cells, defining them as bone-forming cells, below the periosteum which he recognized as cells covering the bone on the surface of Haversian cell system. Carl Gegenbaur, a German anatomist, who, in 1864, defined these cells as osteoblasts. Subsequently, between 1872 and 1873, the functions of osteoblasts and osteoclasts were identified and analyzed.

The latter, in 1873 Kolliker identified osteoclasts as bone resorption cells. However, it was only later in 1849 that Robin differentiated osteoclasts from giant cells of bone marrow but he did not indicate the function of the giant cells of bone marrow.

Robin also described the linear arrangement of osteoblastic cells along the bone wall at the areas of bone formation. He described the inner membrane of the bone, currently known as the endosteum or endosseous surface. We currently know that the bone resorption process is multiphase which was divided, by Vaes, in 1988, into three distinct phases:

1. Formation of osteoclast progenitors from hematopoietic bone marrow, their vascular dissemination to bone surface where they form resting or active osteoclasts.
2. Activation of osteoclasts that are in contact with bone surfaces.
3. Resorption of both organic and inorganic component of bone by the special "ruffled border" with the organic matrix being removed by enzymatic digestion while mineral is solubilized by hydrogenic ion secreted by ATP-driven proton pump in the acidic microenvironment.

In 1910, a group of pathologists and surgeons, including the eminent German pathologist Friedrich von Recklinghausen, definitively established that osteoclasts are the consequence rather than the cause of bone resorption.

1.1.7 Functions of Bone in the Skeleton

The skeleton has several functions that include, support for weight-bearing and locomotion, protection, reservoir for hematopoietic bone marrow, and major reservoir for minerals of the body of vertebrate animals.

The skeleton provides a unique body shape for each of the three sub-categories of mammals, i.e. monotremes, marsupial, and the largest group of placentals. Together there are 27 orders of mammals and further divisions by genus and species. But skeletal shape generally aids in identifying a member of different orders of mammal species.

The skeleton is bilaterally symmetrical and the mid-sagittal plane divides the body into two exactly specular parts. The skeleton occupies a central position within the locomotor system of the organism. Movement, an essential function in every living being, also depends on the action of soft tissues, including muscles, tendons, ligaments, bundles, and the nervous system, but, first of all, it is the bone tissue and joints, which guarantee stability to the various parts of the body. These mechanisms are ensured by the flexion, extension, and rotation of the joints. The entire skeleton is constantly subject to structural changes resulting from the continued growth of body weight. The essential support functions of the skeleton are influenced by the action of various extraskeletal factors. The bone tissue, inserted deep into the soft tissues, is strictly dependent on its vascular and nervous support. Skeletal muscles are inserted into the bone and control its relative movements. Furthermore, bone also plays a central role in contributing to the life support of exchanges and also represents an irreplaceable storage organ, especially for calcium and phosphate. A necessary condition for life is that the serum calcium level is kept at constant levels and, in this case, the contribution that the bone provides is certainly decisive. Under the influence of many hormones, calcium is removed or deposited in the bone according to the needs of the moment. Parathyroid hormone stimulates bone resorption. *Calcitonin*, estrogens, and androgens inhibit bone resorption, *Vitamin D* and somatotropin are required for resorption. *Insulin* for new bone deposition; *Glucocorticoids* inhibit new bone deposition and are also needed for bone resorption. *Tyrosine* stimulates remodeling. *Estrogen* preserves bone tissue. *Androgens* and prostaglandins E2 (PGE2) promote bone remodeling. *TGF-B* is required for deposition ± *and* inhibits bone resorption. *Interleukin* 1 increases during bone formation and inhibits bone resorption. *Interferon* alpha inhibits bone remodeling. In this *manner*, extraskeletal regulatory mechanisms control metabolic exchanges. The availability of calcium and phosphorous is also managed outside of the skeleton, e.g., in the intestine. Abnormalities in these metabolic processes may be sufficient to modify bone structure and shape of the skeleton sufficiently to form diagnostic patterns recognized and diagnosed by radiologists and pathologists.

The alterations of these metabolic processes determine the characteristic transformations of the bone structure whose diagnostic analysis is well interpreted by radiologists and pathologists.

Bone is a mesenchymal tissue of mesodermal origin having a high degree of differentiation. Bone plays two important roles for the body. They are for providing support and participating in physiological functions. Essentially has two main functions: one of support and the other of archiving. Both have, alternatively, a decisive influence on the bone and only under the influence of these physiological functions, the skeleton and its components can play their normal role. Therefore, there is a mutual interaction between the bone structure.

The main mechanical force generated during weight-bearing and locomotion on long bone of an appendicular long bone is compression. Bone is designed have greatest resistance to failure from compression, to a lesser degree under tension and is weakest under forces of torsion. Cortical bone can modify its architecture, i.e. its transverse shape by adding bone to a surface that is experiencing an increase in serial compressive forces from exercise. This is called "cortical drifting." Mechanical forces including a reduction in weight-bearing, aging, and both endocrine and nutritional metabolic effects may provoke increase bone remodeling resulting in bone resorption and the reduction in the numbers of osteons in the cortex. This will cause osteoporosis.

1.1.8 Origin and Dynamic Functions of the Skeleton and their Origins

Under the influence of physiological stimuli, remodeling is more intense in some areas of the skeleton (spine, pelvis, and long bones) than in others (skull).

The cortex is the most stimulated part of the bone and is able to resist pressure and tension forces, responding to their action with physiological changes in shape. In contrast to the high density of this calcified tissue, the (compact) cortex is malleable and this characteristic is maintained by the interpenetration of its Havers canals.

From: The Great Themes of Medicine: The Locomotor System Part I—Fabbri Editore. The skeleton constitutes an organic system assigned to five fundamental functions:

1. Mechanical support of the body;
2. Movement facilitated by muscle, nerve, and voluntary brain integration.
3. Protection and support for external soft tissue and internal organs.
4. Reservoir for mineral stores of the body and a participant in systemic calcium homeostasis for the body.
5. Storage site for hematopoietic tissue that provides the blood supply for the body.

Bone tissue originates in the embryo by one of two developmental mechanisms. Intramembranous bone is formed directly by progenitor mesenchymal cells that arise and fill a fibrous membrane. These osteogenic cells convert the bilayer fibrous membrane into a plate of bone tissue while maintaining the proliferative cells at the surface layer and form a structure termed the periosteum. This development mechanism forms flat bones, e.g. flat bones of the skull plates.

However, most bones of the skeleton are formed of endochondral bone, the second of the two developmental mechanisms. Most of the bones of the axial and appendicular skeleton are formed of endochondral bone, i.e. bones that were initially formed of cartilage by a membrane called perichondrium. The perichondrium had mesenchymal cells that formed cartilage models of the developing bone during fetal development. At a genetically programed time depending upon the location of the bone and species, mesenchymal chondrogenic cells differentiate into osteogenic cells and convert the perichondrium into periosteum. Once formed the periosteum begins to apply layers of bone tissue to the developing cortical surface of the cartilage model transforming the former cartilage model into an endochondral bone. Centrally located cartilage tissue in the developing bone form growth plates or physes. Physes are located at the ends of the developing bone where longitudinally directed interstitial cell division of chondrocytes of the physis provides growth in length of the developing bone model during skeletal development. Simultaneously, osteogenic granulation tissue led by osteoclasts invades the center of the cartilage model of the developing bone to excavate a medullary cavity prior to invading the medullary surface of the physeal cartilage. Bone remodeling units of osteogenic cells led by osteoclasts invade and begin to replace the physeal cartilage with cancellous bone until all of the cartilage is replaced by cancellous bone at time of skeletal maturity. Bone remodeling units consisting of osteoclasts coupled with and followed by osteogenic cells are responsible for shaping the developing bone and remain permanently available to repair injured bone.

1.1.9 Ultrastructure of Cells of the Skeleton

The cell has two major components, the nucleus and the cytoplasm. The first contains DNA of chromosomes in a dispersed state in the nucleus when the cell is not involved in division. During mitosis chromosomes undergo duplication. The nucleus may contain one or more nucleoli. The cell is bounded by a cell membrane.

1.1.10 Cells of the Skeleton Derived from Neural Crest

This subheading suggests that all of the following 13 cells types to be discussed below are of Neural Crest origin which they are not and are therefore confusing since most are formed from mesenchyme derived from mesoderm. My understanding from recent seminars in embryology of domestic animals which was fairly up to date indicates that neural crest cells can form cartilage and bone of craniofacial structures of the head and not of the appendicular and axial skeletal bones, mural cells and pericytes of blood vessels, e.g., and fibroblasts but mesenchyme of mesodermal origin forms most of the tissues of skeleton, i.e. bone, cartilage, blood vessels and lymphatics, skeletal muscle, the connective tissue attachment, fat and associated fibroblasts that form fibrous connective tissue but when irritated can also form fibromyoblasts especially responsible for tissue contraction in wound healing,

1.1.11 Fibroblasts, Fibrocytes, and Myofibroblasts in the Skeleton

Fibroblasts are activated connective tissue cells that synthesize proteins, mainly collagen fibers, reticulin, and elastin and when inactive stain more lightly and are called fibrocytes. Myofibroblasts are fibroblasts that in response to tissue injury acquire contractile stress fibers that aid in wound contraction. While fibroblasts and most myofibroblasts in connective tissue are of mesenchymal mesodermal origin cells, with morphologic features of myofibroblasts are reported to have multiple origins. Myofibroblasts in connective tissue are of mesenchymal mesodermal origin, cells with morphologic features of myofibroblasts are reported to have multiple origins. Their electron microscopic appearance reveals an elliptical nucleus with one or two nucleoli set in cytoplasm containing abundant endoplasmic reticulum often having dilated cisterns. Several mitochondria and a Golgi complex are present.

These cells have the ability to produce numerous substances, including collagen, elastic fibers and reticular fibers, and glycosaminoglycans. Fibroblasts also have retained the capacity to differentiate that allows them to form other connective tissue cells such as chondroblasts, osteoblasts, adipocytes, and smooth muscle cells.

Fibroblasts stain positively to vimentin stain and as they differentiate into myofibroblasts may stain with actin. Myofibroblasts that stain for actin and sometimes myosin have contractile cytoplasmic microfilaments lying adjacent to the cell membrane and that participate in the formation of granulation tissue and in contraction during wound healing as well as the formation of fibrotic tissue in some tumors.

Fibrocytes are mature inactive fibroblasts. They are smaller and more elongated cells that have a dark staining nucleus and a sparse amount of cytoplasm that contains a reduced amount of endoplasmic reticulum.

The most common of the five types of collagens is Type I that makes up 90 percent of the collagen in the body being present in skin, bone, tendon, ligaments, muscle fascia, and supporting connective tissue of the vasculature and organs. Type II forms the collagenous matrix of cartilage. Type III is the main component of reticulin fibers. Type IV collagen is the main fibrous component of basement membranes. Type V collagen contributes formation of fibril formation in other types of collagens.

Collagen fibers form arrays of fibrils that stain mildly positive with the PAS stain, black with silver stains, and a pink color with H&E stain. Electron microscopy investigations have shown that these fibers, as well as others of the connective tissue, have longitudinal streaks and the collagen fibrils, on the other hand, show transverse bands that are the result of overlapping stacks of parallel molecules. The collagen matrix can assume hyaline modifications which give it a homogeneous, vitreous, and rosy appearance (E.E.).

Transmission electron microscopic examination of non-collagenous fibers only displays linear fibers without transverse periodicity. Collagen fibers exhibit transverse periodicity that results from a linear arrangement of sequential fibrillary molecules each one consisting of three collagen polypeptide chains having a triple-helical structure that are unique properties of collagen fibers.

Reticular fibers are a more delicate collagen fibril designated as Type III collagen that is present in basement membranes of epithelial structures, blood vessels, lymphatics as well as minor component of connective tissue that supports many tissues and organs of the body, e.g. nerve sheaths and bone marrow stroma.

Elastic fibers have a natural yellow appearance when viewed by light microscopy and are fluorescent under ultraviolet illumination. Elastic fiber is made of an axial component of longitudinally oriented proteins that provide a positive intrinsic birefringence when viewed in longitudinally oriented sections. They form linear arrays without periodicity when viewed by transmission electron microscopy.

Amorphous matrix, synonyms include extracellular matrix or ground substance, is a gelatinous material consisting of proteoglycans that are formed by the linkage of large molecules of glycosaminoglycans and that contain water while filling the tissue space between connective tissue fibers and cells.

This matrix is markedly basophilic, in preparations stained with Hematoxylin-Eosin, Toluidine Blue, and Alcian blue and is also P.A.S. positive. When there is an excess of water in the matrix causing cellular dispersion and modification in cell shape to spindle and/or stellate forms, this modified appearance is termed myxoid tissue or tissue having undergone a myxoid change.

1.1.12 Monocytes, Macrophages, and Histiocytes

Monocytes are a myeloid hematopoietic cell of bone marrow origin that can enter the circulating blood as monocyte. They can be attracted by inflammatory mediators to leave blood vessels and enter soft tissues where they can transform in to phagocytic cells called histiocytes where they function as macrophages. Other white cells can be phagocytic cells, e.g. neutrophils, mast cells, and dendritic cells but they are not macrophages. Monocytes forming macrophages form the so-called Monocyte-Macrophage system. Therefore, all phagocytic cells are not macrophages. Their main functions are: phagocytosis, antigen presentation to lymphocytes, and recruitment of lymphocytes. Once the macrophage has captured and digested the disease agent; it displays the antigens of the agent on its surface where these antigens are presented to lymphocytes in the lymph node draining the site of inflammation.

Histiocytes originate in the bone marrow and then circulate in the blood as monoliths that can migrate (when activated) to connective tissues where they can be fixed (inactivated) or migrating (activated). Fixed macrophages not unlike inactive fibroblasts can become reactivated and functional. Some connective tissue cells of histiocytic origin and having phagocytic activity have originated from monocytes in bone marrow, entered circulation, and implanted in several tissues. These include Kupffer cells of the liver, mesangial cells of the kidney, alveolar and interstitial histiocytes of the lungs, lymph nodes, spleen, bone marrow, A-cells of the synovium and osteoclasts.

Activated histiocytes, or macrophages, are rather large cells, with abundant cytoplasm, and a central, vesicular nucleus with edges. Their cytoplasm has shaggy extensions with undulating membrane, while performing amoeboid-like movements. The cytoplasm contains lysosomes and phagosomes, microtubules, pinocytosis vesicles, microfilaments and intermediate filaments, and a prominent Golgi complex.

Histiocytes move and engulf dead cells, cell debris, necrotic material, and foreign material. Their cytoplasm is filled with granules and vacuoles, and now this histiocytic cell is performing a phagocytic function and is now regarded as macrophage.

Larger macrophages take on an epithelioid appearance or form to form multinucleated giant cells. From an immunohistochemical point of view, histiocytes and macrophages stain positive for several enzymatic stains, including acid phosphatase, collagenase, lysozyme, alpha-1-chymotrypsin, alpha-1-antichymotrypsin, muramidase, and CD-68.

1.1.13 Chondroblasts and Chondrocytes

When primitive stellate mesenchymal cells become chondroblasts, they lose their cytoplasmic extensions and become round. Their cytoplasm acquires a basophilic appearance due to ribosomal RNA and granularity of the endoplasmic reticulum shared with a prominent Golgi apparatus. Cell density decreases secondary to production and deposition of chondroid matrix by each chondrocyte that results in an expansion of intercellular matrix, i.e. example of interstitial growth.

Chondrocytes are enclosed in a capsule or lacuna that with the immediately adjacent chondroid matrix may stain more basophilic and metachromatically than the interstitial chondroid matrix. Prior to cell division a chondrocyte is oval to round in shape and contains glycogen that can vary in amount. Matrix vesicles released from their cytoplasm initiates calcification of their adjacent matrix. Their capsule can stain positive for the protein S-100 that increases with increased synthetic activity. Intermediate-sized filaments of procollagen and granular material consisting of proteoglycan are observed by transmission electron microscopy. Binucleate chondrocytes in a lacuna indicate cell division that can result in the formation of more cellular aggregates of chondrocytes referred to as "cell clones" or "chondrones."

The intercellular matrix referred to as chondroid matrix is composed mostly of amorphous proteoglycans having an abundance of hyaluronic acid and chondroitin sulfate that bind water and obscure a smaller amount of Type ll collagen fibers when viewed by routine light microscopy. This matrix takes a basophilic stain with H&E, Alcian Blue, and Colloidal Iron stains.

PAS stains glycoproteins a reddish hue while Toluidine Blue is a metachromatic stain.

Cartilage tissue is avascular having no blood vessels or lymphatics; therefore, chondrocytes are nourished by simple diffusion of nutrients through the amorphous matrix. Calcification of cartilage is a normal developmental feature occurring in endochondral ossification where the cartilage model of bones of the axial and appendicular that were initially formed of cartilage are replaced by bone tissue and in growth plates that account for growth in length of bones. Cartilage forms smooth articular surface and buffers forces during locomotion. Following death of the calcified cartilage tissue, those mineralized structures initially formed of hyaline cartilage are removed by multinucleated cells called chondroclasts. Those multinucleated giant cells have the same origin as osteoclasts and function by removing and remodeling mineralized matrix of the former cartilage model. Vascularized osteogenic mesenchyme applies bone tissue to the surface of the calcified model of dead cartilage while that surface is being excavated by chondrocytes. Osteoblasts arising from osteogenic mesenchyme deposit layers of new woven bone tissue on the calcified cartilage surface of the cartilage model and entrap themselves in osteoid matrix while maintaining a layer of osteoblasts lining the new bone surface where they continue to deposit more bone matrix. The new layers of bone matrix of woven bone are composed primarily of Type l collagen fibers and smaller amounts of amorphous matrix. By contrast the cartilage of higher animals, the chondroid matrix of in some vertebrates, fish and amphibians, even in human embryos and in some cartilaginous tumors, their cartilage tissue has greater cellularity and more collagen than amorphous matrix.

1.1.14 Osteoblasts, Osteocytes and Osteoclasts

Osteoprogenitor cells, i.e., preosteoblasts, viewed by H&E stains have an ovoid, polygonal, or spindle shape. The have sparse basophilic cytoplasm, a euchromatic nucleus, and a prominent nucleolus. Ultrastructural images of osteoblasts reveal that the basophilia is caused by rough endoplasmic reticulum and abundant polyribosomes.

A bi-layered membrane called the periosteum arises from mesenchyme and covers bone surfaces but not articular surfaces, linings of tendon sheath sheaths, or surfaces of ligaments. The superficial layer, e.g., fibrous layer of the periosteum, as its name implies has more fibrillary matrix than cells and they resemble fibroblasts.

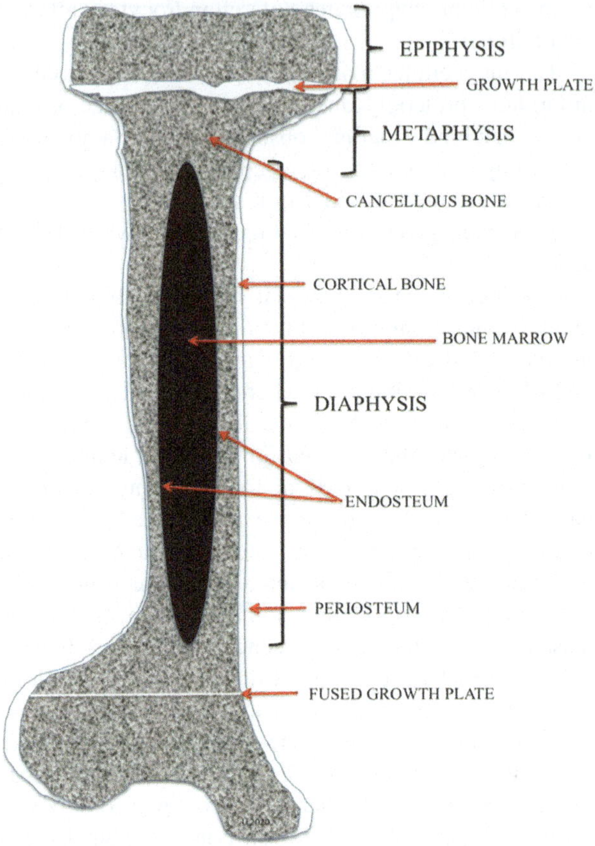

Fig. 1.3 Diagram illustrating structure of normal long bone

However, a few mesenchymal stem cells remain quiescent in that layer and when properly stimulated can participate in formation of the external callus during fracture healing or differentiate into cells that form osteoid and/or cartilage matrix that are features in tumors arising on cortical surfaces of mature bones.

The thicker and deeper layer, i.e., osteogenic layer, is composed of osteoblasts that are in direct contact either with a cartilage surface while they participate in endochondral ossification or with a pre-existing bone surface where they are formed and entrap themselves in bone tissue as osteocytes as they apply layers of bone to a bone surface in a biological process referred to as appositional growth of a bone surface that is undergoing thickening. Mesenchymal stem cells are also preserved in this layer and remain inactive until stimulated to participate in repair, dysplasia or may undergo malignant transformation to form bone tumors. Inactive osteoblasts appear as flat spindle cells on quiescent, i.e. inactive bone surfaces.

Bone is dynamic mineralized tissue (Fig. 1.3) that is constantly being remodeled during the life of the individual with the purpose of renewing microdamage and to create new foci of bone tissue sometimes referred to as multicellular units

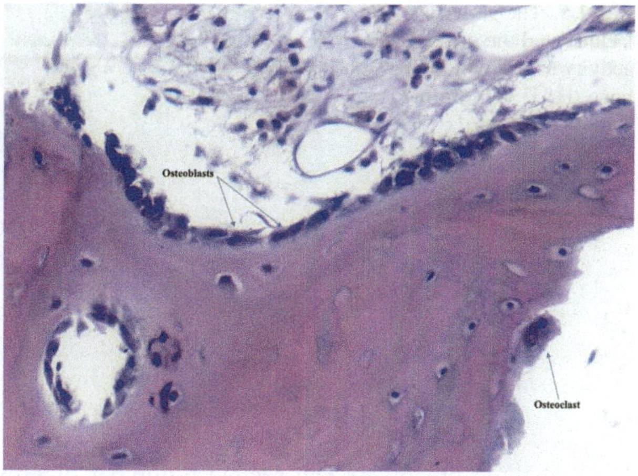

Fig. 1.4 Osteoblast's deposition and osteoclasts demolition in normal physiological bone activities (H&E stain, from author collection)

periodically. This is accomplished by osteoclasts that remove an small areas of bone tissue from a bone surface that osteogenic mechanism refills with viable buried osteocytes that are interconnected by thin cytoplasmic process to one another by gap junctions house in canaliculi in mineralized bone matrix eventually connected to cytoplasmic process in the overlying active or resting osteoblasts lining bone surfaces. Active osteoblast lining cells (Fig. 1.4) can facilitate the flux of calcium ions from bone tissue located in the walls of the canaliculi to the bone surface when the bone lining is stimulated by appropriate hormonal mediated maintenance of calcium homeostasis in the body. Bone remodeling units consisting of a cutting cone of osteoclasts drill cylindrical holes in compact cortical bone along the long axis of a bone to create Haversian canals onto whose walls bone tissue is applied by osteogenic cells that differentiate into osteoblasts in a structure call a filling cone. The completed linear cylindrical bone unit is calling an osteon. Osteons not only participate in calcium homeostasis but in resistance to forces of compression during weight-bearing and locomotion of the skeleton.

Inactive osteoblasts are flat spindle cells, indistinguishable from fibroblasts and fibrocytes.

Active osteoblasts (with an epithelioid appearance) line the surface of the bone or calcified cartilage, producing new layers of bone matrix on the surface. Thus osteoblasts produce their matrix only towards the surface with the pre-existing substrate. This pattern is referred to as "appositional growth" and is opposite to the interstitial growth of cartilage. When a new layer of osteoblasts and bone is added to the surface, the osteoblasts from the previous layer are included in the matrix and turn into osteocytes.

Osteoblasts are joined together and with osteocytes, through tight junctions, also called Gap Junctions. These cellular junction mechanisms also serve as a means of exchange for different types of signal molecules, which regulate cellular metabolic activity and the deposition activity of the bone matrix.

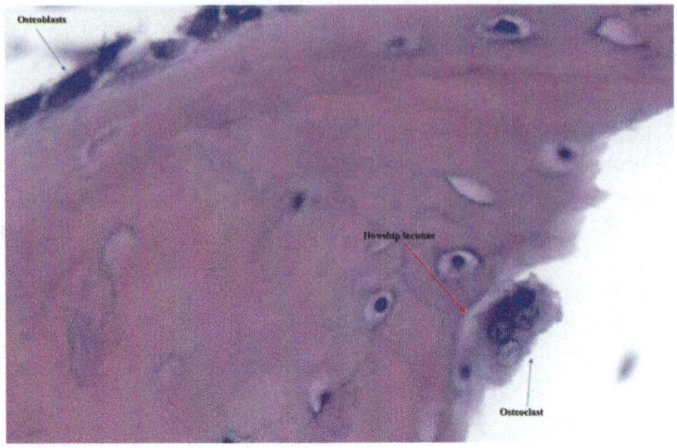

Fig. 1.5 The Howship lacunae and the osteoclastic activity to renew normal bone (H&E stain, from author collection)

Osteogenic precursors can secrete a number of growth and cell differentiating factors and proteins, e.g. bone morphologic protein. Calcium containing intracytoplasmic vesicles released by osteoblasts that are recognized as bone nodules by transmission electron microscopy are morphologic features of the early stages of mineralizing osteoid matrix to form bone tissue. In decalcified bone sections osteoid stains eosinophilic (acidophilic with H&E stains), blue with Mallory's trichrome and positive for PAS because of proteoglycans in the amorphous component of the matrix. However, in undecalcified bone sections bone tissue stains more basophilic, i.e. pink/purple with H&W and red with Mallory's trichrome. Goldner and von Kossa are preferred stains for recognizing osteoid in non-decalcified plastic-embedded bone sections.

Osteoclasts originate from the fusion of cells belonging to the macrophage-monocyte line. They are multinucleated giant cells, with abundant cytoplasm, little basophilic or slightly acidophilic, vacuolized and with a "ruffled border" that when viewed by transmission electron microscopy presents as a complex border of microvilli of the cytoplasmic surface facing and resorbing the bone surface while it forms a crescentic defect called a "Howships lacuna" in the bone surface (Fig. 1.5).

1.1.15 Smooth Muscle Cells

These *spindle-shaped* cells are elongated more than 0.5 mm in length. In the *fetus they have* a fusiform shape, with the nucleus located in the central position of the cell. The nucleus of these cells appears cylindrical, improperly called by some "Havana cigar-shaped." The cytoplasm may appear slender and finely fibrillar with Van Gieson stain. The electron microscope shows a prevalence of rod microfilament *and smaller numbers of* thin myosin filaments *that parallel the long axis*. In addition to these contractile filaments, there are also intermediate filaments of desmin, associated with dense bodies. The Golgi apparatus, the rough endoplasmic reticulum, and free ribosomes *that* are mainly distributed near to the poles of the nucleus.

The cytoplasm contains moderate amounts of glycogen and is positive, immunohistochemically, for desmin, actin, and vimentin. The cells aggregate in bundles and around each cell it is possible to detect a delicate basement membrane containing reticular fibers. Smooth muscle cells normally synthesize and secrete collagen, elastin, proteoglycans and the basement membrane that surrounds them.

1.1.16 Striated Skeletal Muscle Cells

Mature skeletal muscle cells are derived from mesenchyme. They range in length from 1–40 mm and range from 10–50 mm in width. They have a cylindrical shape and a transversely striated cytoplasm. They may contain as many as 100 nuclei.

The cytoplasm is formed primarily of parallel aggregates of myofibrils. Each myofibril consists of myofilaments that produce a histologic pattern of alternating light and dark bands of myofibrils throughout the length of the muscle fiber. Thin and thick myofilaments, respectively, of actin and myosin of myofibrils can be viewed by electron microscopy. These muscle fibers have a scant amount of cytoplasm, i.e. sarcoplasm that contains smooth endoplasmic reticulum, a small Golgi complex ribosomes, and glycogen. Nuclei are periodically aligned on the inner surface of the cell's plasmalemma named in this cell, the sarcolemma.

The nuclei are flattened and located at the periphery, below the sarcolemmal membrane. A basement membrane is applied to the outer surface of the sarcolemma and a mesenchymal stem cell called a "myosatellite cell" is incorporated within the sarcolemma to participate in tissue repair following muscle damage. When it is called upon to initiate muscle fiber repair, it passes into the sarcoplasm and differentiates into a myoblast. Chains of myoblasts align themselves in single file, fuse their cytoplasm with neighboring myofibroblasts to form a chain that can regenerate and repair the damaged striated muscle fiber. Immunochemical stains for striated muscle fibers and myoblasts are positive for desmin, actin, and myoglobin.

1.1.17 Lipoblasts and Lipocytes

When the primitive mesenchymal cells are transformed into lipoblasts, modifications are observed essentially characterized by the presence of small cytoplasmic lipid droplets, arranged around the central nucleus. With cellular maturation, the lipid vacuoles merge into a single large vacuole (liposome). The mature cell appears large, spherical, or polyhedral, and the nucleus moves towards the periphery. The lipid content of lipoblasts and lipocytes dissolves in histological preparations, treated with alcohol, but can be preserved in the sections obtained by freezing fresh samples. The lipids are highlighted with histochemical colors such as Sudan III, Sudan black, Nile Blue, and Congo Red. They can also be preserved and colored with osmic acid. Lipocytes are immunohistochemically positive for the S-100 protein.

During embryonic development, adipose tissue (white fat) grows with a lobular architecture and is accompanied by a delicate proliferation of a capillary network.

1.1.18 Angioblasts, Vascular Endothelium, and Pericytes

The vessels originate from primitive angioblasts: large cells, with P.A.S. positive, which becomes vacuolized. The vacuoles widen, causing a displacement of the nucleus to the periphery of the cell, deform in the shape of a crescent, and begin to give shape to what appears as a primitive sketch of a vessel lumen. With proliferation some primitive endothelial cells line the vessel and appear as enlarged cells around a very narrow lumen. As the vessels mature, the endothelial cells arrange themselves in a thin, flattened layer that enlarges the vessel. Endothelial cells with their flattened cytoplasm are connected to each other by a system of intercellular junctions that determine the formation of a real cell wall. The endothelial cells are placed on a basement membrane that contains reticular fibers and P.A.S. positive in its extraluminal surface. Both surfaces, luminal and extraluminal, are disseminated by small pinocytotic vesicles which serve to transport fluids within the cells.

Ultrastructurally, endothelial cells exhibit “rod body” structures containing parallel tubules (Weibel–Palade bodies). These bodies appear to contain the von Willebrand factor (vWF), responsible for platelet action.

Endothelial cells are immunohistochemically positive for factor VIII, CD31, and CD34.

The capillaries and venules are occasionally surrounded by pericytes, cells with an elongated nucleus, and an extensive cytoplasm. Pericytes are cells often associated with the endothelium. They have a stem-cell character because after a capillary break they seem to be able to regenerate damaged tissue and, moreover, they appear as precursors of fibroblasts (http://www.lacellula.net/appunti/istologia/componale_cellulare_tessuto_connettivo.html). They are immunohistochemically positive to Vimentin and factor VIII and contain microfilaments. They also have contractile properties and can also perform macrophage-like functions.

Ultrastructural examination finds “rod-like” structures containing parallel tubules called Weibel–Palade bodies that are said to contain von Willebrand factor. This large glycoprotein has critical functions in plugging defects in vessel wall to control hemorrhage. Immunohistochemical stains for endothelial cells include Factor VIII, CD31, and C34. Primitive mesenchymal stem cell precursors of endothelial cells remain in a quiescent stage among the endothelial lining cells where they are important in vessels repair and budding off of new capillaries from vascular lumens. The basement membrane of the vascular wall also contains primitive mesenchymal stem cells called pericytes. They also can respond in the repair of vascular damage but they also have the potential to differentiate into along other mesenchymal tissue lines in pathologic disorders. Pericytes contain microfilaments, exhibit contractile features, and/or demonstrate phagocytic behavior under certain conditions. They histochemically stain for Vimentin and Factor VIII.

1.1.19 Synovium

The normal synovial membrane essentially consists of two types of cells (synoviocytes): type A or macrophage-like and type B or fibroblast-like. Type A cells line the synovial surface, similar to an epithelium, but form an incomplete layer and do not rest on a basement membrane. These arise from mononuclear cells produced in the bone marrow. Type B cells are more numerous, resemble fibroblasts, and are of local or intrinsic, i.e. mesenchymal origin. Other hypotheses consider both histotypes as different functional expression of the same origin. Ultrastructurally, Type A cells are characterized by long cytoplasmic processes, perpendicularly directed towards the joint cavity, i.e., filapodia, but do not have complex junctions or a basement membrane.

Immunohistochemically they are positive for Vimentin and negative for cytokeratins and epithelial membrane antigen. Synoviocytes synthesize hyaluronates, sulfurized glycosaminoglycans, and binding proteins. Both types of synovial cells, A and B, may occasionally express monocyte antigens. Both histotypes have powerful phagocytotic properties, but A cells are much more active than B cells. The synovial membrane is rich in blood and lymphatic vessels.

1.1.20 Mast Cells

These cells are derived from hematopoietic cells in the bone marrow and are widely distributed in the body.

They are generally dispersed in connective tissue, around small arteries and their neighboring bundles of nerves. Numerous granules in their cytoplasm are coated with a membrane that takes a metachromatic stain, e.g., Toluidine blue or Giemsa.

The granules contain different categories of substances such as histamine, heparin, chondroitin sulfate, and various proteolytic enzymes. Mast cells play a fundamental role in some humoral and cell mediated immune mechanisms (hypersensitivity and anaphylaxis).

1.1.21 Cells of Neural Crest Origin

The neural crest, in the embryo, is located lateral to the neural tube. Of ectodermal origin, the neural crest gives rise to epidermal melanocytes (and is also capable of epithelial differentiation), ganglia and sensory fibers of the spinal nerves (somatic), ganglia and nerves of the sympathetic (autonomic) system, the medulla of the adrenal gland and the chromaffin cell system and Schwann cells. In the peripheral nervous system the neuronal cells of the autonomic and somatic ganglia are large and round in shape, with a vesicular nucleus (the sympathetic ganglion cells are often binucleated), with prominent nucleoli. The cytoplasm contains a typically basophilic substance, arranged in clumps of Tigroid or Nissl substance, mainly consisting of rough endoplasmic reticulum and free ribosomes. The cytoplasm contains Golgi apparatus, mitochondria, lysosomes, neurofilaments and neurotubules and Actin microfilaments. Neurofilaments and neurotubules extend from the cell body to the axon. Immunohistochemically, neuroganglionic cells stains for neuron-specific enolase, neurofilament protein, synaptophysin, and chromogranin.

The axon of the cells (nerve fiber) is lined with Schwann cells, which have an elongated nucleus, in a lamellar cytoplasm. When the lining of the axon is made up of a single layer of Schwann cells, the nerve fiber is unmyelinated, as occurs in sympathetic nerve fibers. In myelinated nerve fibers, however, the cytoplasm of Schwann cells produces a multilayered coating. These concentric layers of the cytoplasmic membrane of Schwann cells constitute a true myelin sheath. Schwann cells positively stain with the S-100 protein. Together with Schwann cells, the peripheral nerve sheaths are composed of fibroblasts and collagen (endoneurium, perineurium, and epineurium) which have a mesenchymal derivation.

Chromic salts (chromaffin) are used to stain cells of the adrenal gland for light microscopy cells. Viewed by transmission electron microscopy their cytoplasm contains large quantities of electron dense granules that contain adrenalin and noradrenalin (Hall 2005; Doussis et al. 1992; Jubb et al. 2007; Nielsen and Rosenberg 2012; Szymendera 1970; Zelzer and Olsen 2003; Campanacci 1999).

References

Adler C-P (2000) Bone diseases: macroscopic, histological, and radiological diagnosis of structural changes in the skeleton, 1st edn. Springer, London

Akatsu T, Takahashi N, Udagawa N, Sato K, Nagata N, Moseley JM, Martin TJ, Suda T (1989) Parathyroid hormone (PTH)-related protein is a potent stimulator of osteoclast-like multinucleated cell formation to the same extent as PTH in mouse marrow cultures. Endocrinology 125:20–27

Bani D. (2007) Il tessuto osseo. http://www.rsmilano.it/1/PDF/Il%20Tessuto%20Osseo-ProfDanieleBani.pdf

Baron R, Neff L, Van P, Nefussi JR, Vignery A (1986) Kinetic and cytochemical identification of osteoclast precursors and their differentiation into multinucleated osteoclasts. Am J Pathol 122: 363–378

Campanacci M (1999) Bone and soft tissue tumours: clinical features, imaging, pathology and treatment, 2nd edn. Springer, Vienna

Doussis I, Puddle B, Athanasou NA (1992) Immunophenotype of multinucleated and mononuclear cells in giant cell lesion of bone and soft tissue. J Clin Pathol 45:398–403

Farnum CE, Wilsman NJ (2002) Chondrocyte kinetics in the growth plate. In: Shapiro IM, Anderson HC, Boyan BD (eds) The growth plate. IOS Press, Amsterdam, pp 245–257

Farnum CE, Lee R, O'Hara K, Urban JPG (2002) Volume increase in growth plate chondrocytes during Hyperthrophy: the contribution of organic Osmolytes. Bone 30:574–581

Farnum CE, Lee AO, O'Hara K, Wilsman NJ (2003) Effects of short-term fasting on bone elongation rates: an analysis of catch-up growth in young male rats. Pediatr Res 53:2003

Gorski JP, Griffin D, Dudley G, Stanford C, Thomas R, Huang C, Lai E, Karr B, Solursh M (1990) Bone acidic glycoprotein-75 is a major synthetic product of osteoblastic cells and localized as 75-and/or 50-kDa forms in mineralized phases of bone and growth plate and in serum. J Biol Chem 265:14956–14963
Hall BK (2005) Bones and cartilage: developmental and evolutionary skeletal biology. Elsevier-Academic Press, Burlington
Horton MA, Rimmer EF, Lewis D, Pringle JA, Fuller K, Chambers TJ (1984) Cell surface characterization of the human osteoclast: phenotypic relationship to other bone marrow derived cell types. J Pathol 144:281–294
Jubb KVF, Kennedy PC, Palmer N (2007) Pathology of domestic animals: chapter 1 – bone and joints, vol 1, 5th edn. Elsevier Saunders, New York, pp 119–120
Karsenty G (2003) The complexities of skeletal biology. Nature 423:316–318
Kronenberg HM (2003) Developmental regulation of the growth plate. Nature 423:332–336
McSheehy PMJ, Chambers TJ (1986) Osteoblastic cells mediate osteoclastic responsiveness to parathyroid hormone. Endocrinology 118:824–828
Nielsen GP, Rosenberg AE (2012) Diagnostic pathology bone, 1st edn. Amirsys Inc., Philadelphia, PA
Nobta M, Tsukazaki T, Shibata Y, Xin C, Moriishi T, Sakano S, Shindo H, Yamaguchi A (2005) Critical regulation of bone morphogenetic protein-induced osteoblastic differentiation by Delta1/Jagged1-activated notch 1 signaling. J Bone Chem 280:15842–15848
Quinn JM, McGee JO, Athanasou NA (1994) Cellular and hormonal factors influencing monocyte differentiation to osteoclastic bone-resorbing cells. Endocrinology 134:2416–2423
Schuenke M, Schulte E et al (2007) Head: cranial bones. In: Thieme atlas of anatomy: head and neuroanatomy, 1st edn. Georg Thieme Verlag, New York
Shinar DM, Rodan GA (1990) Biphasic effects of transforming growth factor-beta on the production of osteoclast-like cells in mouse bone marrow cultures: the role of prostaglandins in the generation of these cells. Endocrinology 126:3153–3158
Suda T, Takahashi N, Martin TJ (1992) Modulation of osteoclast differentiation. Endocrine Rev 13: 66–80
Suda T, Udagawa N, Nakamura I, Miyaura C, Takahashi N (1995) Modulation of osteoclast differentiation by local factors. Bone 17(Suppl 2):87S–91S
Szymendera J (1970) Bone mineral metabolism in cancer, 1th edn. Springer, Berlin
Takayanagi H, Kim S, Koga T, Nishina H, Isshiki M, Yoshida H, Saiura A, Isobe M, Yokochi T, Inoue J, Wagner EF, Mak TW, Kodama T, Taniguchi T (2002) Induction and activation of the transcription factor NFATc1 (NFAT2) integrate RANKL signaling in terminal differentiation of osteoclasts. Dev Cell 3:889–901
Tezuka K, Yasuda M, Watanabe N, Morimura N, Kuroda K, Miyatani S, Hozumi N (2002) Stimulation of osteoblastic cell differentiation by notch. J Bone Miner Res 17:231–239
Udagawa N, Takahashi N, Akatsu T, Tanaka H, Sasaki T, Nishihari T, Kosa T, Martin TJ, Suda T (1990) Origin of osteoclasts: mature monocytes and macrophages are capable of differentiating into osteoclasts under a suitable microenvironment prepared by bone marrow derived stromal cells. Proc Natl Acad Sci U S A 87:7260–7264
Zelzer E, Olsen BR (2003) The genetic basis for skeletal diseases. Nature 423(6937):343–348

Chapter 2
Tumors of the Musculoskeletal System

2.1 Terminology

2.1.1 *Hyperplasia*

This term refers to an increase in the size of a tissue or organ due to an increase in the number of cells and their cell products due to mitotic activity. Hyperplasia in a normally formed tissue or organ is a functional response to injury, e.g. callus formation of a fractured bone, to increased hormonal secretion, e.g., pregnancy associated changes including lactation, stress-inducing cortical hyperplasia of the adrenal gland that increases ACTH levels and its secondary effects.

Many cases of functional hyperplasia can mimic pathologic enlargements of organs caused by neoplasia. However, most enlargements caused by functional hyperplasia and are physiological and/or mechanical response to stimuli that may regress spontaneously without progressive enlargements caused by neoplasia; however, the cause of the stimulus for functional hyperplasia may not be readily apparent at the time of clinical observation of the enlargement, e.g. lesions of nodular hyperplasia may be difficult to distinguish from hepatoma while biliary epithelial hyperplasia in rabbits can be demonstrated to have been caused by Eimeria stiedae.

Osteoscleroses An increase of density of internal structures of bone determined by an increasing of mineralization of the tissue and deposition of connective fibrous tissue and caused from different causes that can appear in the skeleton or in single bones involving frequently also the bone marrow (osteomyelosclerosis) (Campanacci 1999; Adler 2000; Dorfman and Czerniak 1998; Fletcher et al. 2013; Inwards and Unni 1995; Jacobsen 1971; Jaffe 1958; Jubb et al. 2007; Layfield 2009; Lichtenstein 1972; Lopez 2011; Marconato and Del Piero 2005; Meuten 2020; Moss and Blser 2005; Nielsen and Rosenberg 2012; Resnick et al. 1995; Rosenberg 2010; Slauson and Cooper 2002; Wrigley 2000; McGavin and Zachary 2007; Pool 1990; Willis 1967).

L. Leonardi, *Bone Tumors in Domestic Animals*,
https://doi.org/10.1007/978-3-030-90210-0_2

2.1.2 *Dysplasia (and Hamartoma, and Choristoma)*

These generic terms indicate an abnormal tumor-like overgrowth of a normal or abnormal tissue. Dysplasia is a non-neoplastic lesion and this is cured by removal. In bone pathology the term dysplasia is frequently used to describe a specific lesion, i.e. hip dysplasia or osteochondrodysplasia. Skeletal dysplasia in man is discussed in Classification of Skeletal Dysplasia, authored by J. Spranger and L.O. Langer and H.R. Widemann, 1974. Aristotle introduced the term, Hamartia, in "Poetics" as an error or mistake. That term is the origin of the term Hamartoma which is used in medicine for a developmental disturbance characterized as an abnormal overgrowth of otherwise normal tissue in a normal location, e.g., a birthmark on the skin. By contrast a Choristoma is a developmental lesion of normal tissue that arises in an abnormal location, e.g. nodule of exocrine pancreatic tissue on the surface of the small intestine. Both abnormalities have histologic features of their respective tissue of origin. Suggested examples of these growth abnormalities are some types of exostoses, chondromas, fibrodysplasia, and chondrodysplasia. While most are solitary lesions, some can be multicentric.

2.2 Abnormal Growths Discussed vs Skeletal Tumors

Tumors While Erasmus, who was perhaps the greatest scholar in the Northern Renaissance period, was quoted as having said, "that it is impossible to define a tumor." Also, if in 1925 in Erasmus Wilson Lectures, G. W. Nicholson asserted "that is impossible to define a tumor," there are several definitions to describe a tumor but we like to consider to frame a satisfactory definition as like as did by K. A. Willis who defined: "a tumor is an abnormal mass of tissue, the growth of which exceeds and is uncoordinated with that of the normal tissues, and persists in the same excessive manner after cessation of the stimuli which evoked the change." Willis described the abnormal mass where "the essential tissue of a tumour is an actively growing cells of a specific kind or kinds which comprise the tumour proper, as distinct from incidental cells of other kinds which from the stroma of the growth or which are due to secondary inflammatory or phagocytic reactions in it. In most tumors the neoplastic tissue consists of cells of a single kind derived from a single kind of tissue and where uncoordinated cellular proliferation, in excess of that of the normal tissues, is a main characteristic of neoplasia, distinguishing it from all other kinds of pathological proliferations, whether inflammatory, reparative or hyperplastic." The same Author reported that "usually, once a tumour is engendered, its cells continue indefinitely to outgrowth of those cells of the normal tissue (excessive growth persisting after cessation of the stimuli which evoked the change)." As reported in WHO Histological Classification of Bone and Joint Tumors of Domestic Animals: "the accurate diagnosis of bone and joint tumors requires adequate clinical and radiographic evaluations of the lesions as well as histology. Without this information the significance of the microscopic findings in a small biopsy specimen may be misinterpreted. A definitive diagnosis may require multiple biopsies." As also reported in his 1976 essay by E. Grundmann: "the diagnosis of a bone tumor has to be based not only on histology, but also on the roentgenogram. If the pathologist is not well trained in the interpretation of Roentgenograms—and this may be also normal—close cooperation between X-ray specialist and pathologist is a prerequisite of effective and detailed diagnosis. In many cases cytologic methods will facilitate, accelerate, or verify the results of diagnosis (Thompson et al. 2016; Willis 1967; Berzina et al. 2008). An exact and detailed diagnosis is the prerequisite of adequate therapy."

2.3 Benign vs Malignant Bone Tumors

BENIGN TUMOR: A "tumor" by definition is a swelling of a part of the body, generally without inflammation, caused by an abnormal growth of tissue, whether benign or malignant. A benign neoplastic mass is referred to a benign tumor to distinguish it from a malignant or cancerous tumor. A benign tumor can grow and increase in size, but it will not spread diffusely. Its borders are fairly well delineated and it does not metastasize. By comparison a malignant tumor, also called a cancer, can also grow, widely infiltrate and increase in size; however, its borders are obscure, i.e., poorly defined due to infiltration and differs from a benign tumor because of its ability to spread widely and metastasize. Confusion between these two extremes in biological behavior can be caused by some benign bone tumors of that are locally aggressive, provoke osteoclastic mediated bone destruction but rarely metastasize to the lung, e.g. $<2\%$ of giant cell tumor of bone.

- The binomial classification of tumors considering the histogenetic origin of neoplasms divided its in epithelial with -oma suffix (e.g., papilloma) and in mesenchymal with -oma suffix (e.g., osteoma). Examples: osteoma, chondroma, osteochondroma, ossifying fibroma, hemangioma, myxoma of the jaw, feline osteochondromatosis, etc.

MALIGNANT TUMOR: As previously indicated a malignant tumor is also called a malignant neoplasm and a cancer. The often-rapid growth, permeative infiltrative spread and destruction of the pre-existing tissue architecture of its site of origin, local vascular invasion, and metastatic behavior characterize this tumor. A binomial classification is also applied to identify malignant tumors is based upon their origin from two developmental tissues: epithelial or mesenchymal. Nomenclature for malignant tumors of epithelial origin is the use of the suffix, −carcinoma, e.g., squamous cell carcinoma, while for malignant tumors of mesenchymal origin by use of the suffix, −sarcoma, e.g., osteosarcoma, chondrosarcoma, fibrosarcoma, liposarcoma. Because of differences in malignant biological behavior, sarcomas are further classified as to anticipated Low-grade and High-grade malignancy. Dr. Mario Campanacci (1932–1999), Director of first Orthopedic Clinic at Rizzoli Human

Orthopedic Institute in Bologna, Italy reported in his textbook, "*these definitions are schematic and approximate. At times it is difficult to define the limits between hyperplasia, hamartoma, dysplasia and benign tumor. A hamartoma or dysplasia may, even after many years, constitute the site where malignant tumors originate. (. . .) There are benign lesions which grow rapidly or which have permeative growth. On the contrary, some malignant tumors are characterized by slow growth; others remain small and clearly circumscribed or have an organoid structure.*"

What Claus-Peter Adler reported in the Bone Diseases textbook in 2000 is also very informative: *the pathologist who has the task of finding the "definitive" diagnosis, must above all have the precise knowledge of the macroscopic and microscopic morphology, and must be able to assign each tumor to its correct place in the generally accepted scheme of classification.* (Adler 2000; Brodey et al. 1963; Cooley and Waters 1997; Dernell et al. 2006; Goldschmidt and Hendrick 2002).

2.4 Cellular Features: Metaplasia and Anaplasia

- METAPLASIA: This term is used to describe a change in a cell form, structure, and appearance that is not normal to the tissue in which it is located. There is evidence that a mature cell and/or usually an accompaniment of cells in a mature tissue or part of an organ has attempted to adapt to chronic changing environmental stimuli in an attempt to survive. Examples for epithelial surfaces would be conversion of ciliated respiratory epithelial cells in pulmonary airways into squamous epithelium and for mesenchymal derived tissue such as tendinocytes, in athletic horses in which chronic tensile stress and ischemia cause the tendinocytes to die or undergo metaplasia to chondrocytes.
- ANAPLASIA: This term is used to describe mature cells or tissues that are derived either of epithelial or mesenchymal origin, which have lost their mature histologic architecture, specialized anatomic features, normal physiological functions and their orientation and normal interaction with each other that caused them to be recognized as being different from their peers. Anaplasia is an important preneoplastic indicator indicative of early malignant change. Atypia is a term frequently used to indicate the presence of cells with anaplastic features in the diagnostic pathology reports. Anisocytosis, i.e., variation in cell size and anisokaryosis, i.e., variation from normal nuclear size are terms used in the descriptions of anaplasia (Campanacci 1999).

2.5 Tumor Progression: Last Phase in Tumor Development

This term describes continuous growth and invasiveness of tumor that are influenced by many factors, e.g., genetic in origin, tissue-mediation, and hormonal-driven, that affect the phenotypical appearance and biological behavior of tumor cells. These factors promote the formation of cellular clones having a high degree of malignancy and metastatic potential (Grundmann 1976; Knecht and Priester 1978; Liu et al. 1974; Liu et al. 1977; Slayter et al. 1994).

2.6 Predisposing Conditions and Genetic Predisposition

While bone tumors appear to arise de novo, it is evident that the cell of origin for many bone tumors is from mesenchymal stem cells present among the stromal cells of bone marrow. Mutations in mesenchymal stromal cells may result from exposure to radiation or cancer-causing chemicals and viral infections, but most are spontaneous and occur for no apparent reason. These mutations are present only in the cancer cells, so they cannot be passed on to offspring. Both in humans and animals some bone tumors can also arise in non-neoplastic lesions of bone where different factors can act as initiators and promotes of the neoplastic transformation. Examples of predisposing factors for bone tumors in animals and man include: radiotherapy, chemotherapy, bone infarction, chronic osteomyelitis, metallic implants, and malignant transformation in a benign bone tumor. In man Paget's disease, fibrous dysplasia and familial retinoblastoma are additional predisposing causes of bone tumor development. The author and a team of colleagues have recently published an investigation of microRNAs as to their roles in spontaneous development and progression of osteosarcoma in man and animals to try to understand better the genetic and biomolecular mechanisms that initiate spontaneous bone cancer. This knowledge should aid them in understanding the etiology and pathogenic mechanisms that are concerned with the origin, maintenance, and malignant progression of osteosarcoma and other bone tumors. This will enable them to develop better approaches for controlling and eliminating these tumors (Thompson and Pool 2002; Jacobsen 1971).

2.7 Tumor Site of Origin Defines Tumor Categories for Osteosarcoma

2.7.1 Periosteal Origin

As used to describe the location of a bone lesion means a lesion that develops in the part of the diaphysis of the bone where is present the periosteum.

2.7.2 Parosteal (SYN.) Juxtacortical

Are terms used to describe the location of the origin of a bone tumor arising on a bone surface but still located within the periosteum. A parosteal osteosarcoma arises by neoplastic transformation of mesenchymal stem cells that has been left behind during limb development within the fibrous layer of the bi-layered periosteum. Parosteal osteosarcoma does not arise from mesenchymal stem cells of fibrous tissues attached to the outer surface of the fibrous periosteum as once incorrectly thought. This is a tumor initially of low-grade malignancy. It can produce seemingly orderly structural elements formed of osteoid, immature woven bone, and some chondroid matrix sometimes making diagnosis from a bone needle biopsy difficult. This is not surprising since benign mesenchymal stem cell counterparts proliferate following injury and form the external callus in fracture repair which contains similar matrix components. Some of these initially low-grade parosteal osteosarcomas can undergo malignant progression to become a dedifferentiated parosteal osteosarcoma. They can then invade the cortex, enter the medullary cavity and, if not removed, may in time metastasize (Dahlin and Unni 1977; Hammer et al. 1995; Kirpensteijn et al. 2002; Kundu 2014; LaRue et al. 1986; Silver et al. 2001; Vanel et al. 2012; Vlychou and Athanasou 2008).

2.7.3 Medullary Origin: Intramedullary (Syn. Intracortical or Central) Bone Tumors

Are terms that designate tumors which arise in the medullary cavity of a bone.

A conceptual approach is offered here to help anyone who is initially attempting to understand the complex classification nomenclature system for benign and malignant tumors of bone. The current classification for human and animal bone tumors is based upon the anatomical location, morphologic appearance, and clinical behavior, i.e. benign behavior, malignant progression that may occur in some benign tumors and malignant behavior from inception of a tumor. It is becoming increasingly clear that most tumors in epidermal and mesodermal derived tissues arise from stem cells left behind as reserve repair cells during fetal tissue development.

Skeletal tumors originate primarily from mesenchymal stem cells (MSCs) that are provoked into undergoing neoplastic transformation by multiple stimuli. While in their initial stage of development, MSCs have pleomorphic potential. However, their range of cellular differentiation becomes limited by the end of the fetal period relative to their location in unique anatomical structures in which they resided. For example, those mesenchymal stem cells that became located in subcutaneous adipose tissue, fascia, tendon sheaths, periosteum, endosteum, and marrow adipose tissue had limitations in the range of morphologic cell types that they can produce. However, it has become apparent that some of these differentiated MSCs can revert to the pleomorphic state. Using this conceptual approach, it becomes easier for students to master the classification of skeletal tumors.

Periosteal tumors of bone: MSCs present in the outer fibrous layer of the periosteum produce cancellous and dense osteomas or periosteal chondromas. MSCs present in the deeper layer of the fibrous periosteum can form periosteal chondrosarcoma and parosteal osteosarcoma while MSCs in the osteogenic layer of the periosteum can form low- and high-grade periosteal osteosarcomas and surface osteosarcomas.

Central tumors of bone: MSCs present among the fat cells and bone marrow stromal cells can form an osteosarcoma, i.e., non-productive or productive osteosarcoma, or the several mixed patterns of osteosarcoma, e.g. chondroblastic, fibroblastic, telangiectatic, and giant cell variants. Their diagnosis is based upon the major pattern in the tumor although that pattern can vary in different areas of the same specimen indicating the pleomorphic capacity of MSCs. MSCs of the marrow can produce one of the other patterns of central bone tumors, i.e. chondrosarcoma, fibrosarcoma, liposarcoma, and giant cell tumor. Central hematopoietic tumors arise from hematopoietic stem cells located in the so-called niches in the medullary cavity while central hemangiomas and hemangiosarcomas arise from pre-existing medullary vessels. Histologic examination reveals that a telangiectatic osteosarcoma is in reality a non-productive osteosarcoma that develops rents in the tumor stroma that fill with blood. This concept of tumor origin and behavior reflects the Medieval premise that you are destined to perform in whatever status of life into which you are born and are destined to perform that function unless you can rise above your status in life, e.g. distinguish yourself by undergo dedifferentiation.

2.8 Introduction to Tumors of Bone

The World Health Organization International Histological Classification of Bone and Joint Tumors of Domestic Animals of 1994, classifies medullary bone tumors of animals into two major groups based upon histologic features that reflect and predict their biological behavior, i.e., benign tumors and malignant tumors. Primary medullary bone tumors are then classified also according to the most prevalent type of neoplastic cells, i.e., (1) resembling the cell of origin: osteogenic, chondrogenic, fibroblastic, giant cells, lipoblasts, angiogenic and (2) matrix, e.g. osteoid, chondroid, fibrous, adipose, etc. The confusion that occurs with the several subtypes of osteosarcoma has become more understandable as a result of stem cell research. These osteogenic tumor subtypes are the progeny of a malignant bone marrow stromal stem cell that can make either a consistent clone of, e.g., osteoblastic osteosarcoma, chondroblastic osteosarcoma, fibroblastic osteosarcoma, giant cell osteosarcoma, or liposarcoma or generate neoplastic clones that produce variable amounts of matrix in different regions of a bone tumor where more than one subtype is present but only one type is represented in a single biopsy specimen taken from that tumor. This has clarified the differences observed in bone needle biopsies from different locations in a bone tumor all being from an osteosarcoma.

Since the 1994 WHO animal bone tumor classification was published, the 2013 World Health Organization Classification of Tumors of Soft Tissue and Bone (Man) Vol 120, Issue 12, June 15, 2014 pp. 1763–1774 recognizes in man two additional tumors: 1) a low-grade sarcoma of bone termed a "Pleomorphic sarcoma" and 2) a high-grade sarcoma of bone termed an "Undifferentiated high-grade pleomorphic sarcoma." Previously, both of these uncommon tumors had been included as osseous examples of malignant fibrous histiocytoma. That tumor designation includes a spectrum of soft tissue tumors of muscles and tendons having somewhat similar histologic features as these two types of pleomorphic sarcomas of bone. However, these two newly recognized pleomorphic tumors like other medullary bone tumors arise from bone marrow mesenchymal stem cells and not from mesenchymal stem cells in extraosseous tissue. In busy veterinary diagnostic services, most veterinary pathologists are not aware of these two types of rare bone tumors and diagnose them as spindle cell sarcomas of bone while remarking upon their absence of matrix production and lower grade malignancy than that of a conventional osteosarcoma or fibrosarcoma.

The general evaluation parameters of bone tumors will be treated in more detail in the following chapters of this text, so please refer to the specific reading in the thematic sections of the book.

To reach a definitive histopathologic diagnosis one is advised by the WHO International Collaborating Center for Comparative Oncology in collaboration with the US Armed Forces Institute of Pathology in Washington, DC that, "the accurate diagnosis of bone and joint tumors requires adequate clinical and radiographic evaluations of the lesion as well as histology," because one biopsy specimen may not be representative of the type and grade of the tumor. One most also reflect on the comments of a pioneer medical pathologist, H.L. Jaffee, who in 1958 commented, "*each tumor must be considered as an anatomical and clinical entity in itself.*"

Primary bone tumors are common in dogs and cats than in other domestic and non-domestic animals.

2.9 WHO Bone Tumors Classification Scheme for Domestic Animal Bone Tumors

The histological WHO 1994 classification of bone and joint tumor of domestic animals comprise:

I. **BENIGN TUMORS**:

A. Osteoma (and osteoblastoma not classified in WHO).
B. Ossifying fibroma.
C. Myxoma of the jaw.
D. Osteochondroma.
E. Feline osteochondromatosis.
F. Chondroma.
G. Hemangioma.

II. **MALIGNANT TUMORS.**

A. Central.

1. Osteosarcoma.

(a) Poorly differentiated.
(b) Osteoblastic.

(i). Non-productive.
(ii). Productive.

(c) Chondroblastic.
(d) Fibroblastic.
(e) Telangiectatic.
(f) Giant cell type.

2. Chondrosarcoma.
3. Fibrosarcoma.
4. Hemangiosarcoma.
5. Giant cell tumor of bone.
6. Multilobular tumor of bone.
7. Liposarcoma.
8. Myeloma.
9. Malignant mesenchymoma.

B. Peripheral.

1. Periosteal chondrosarcoma.
2. Periosteal fibrosarcoma.
3. Maxillary fibrosarcoma (dogs).
4. Periosteal osteosarcoma.
5. Parosteal osteosarcoma.

C. Joint Tumors.

1. Synovial sarcoma.

D. Miscellaneous tumors.
   1. Liposarcoma.
   2. Malignant mesenchymoma.
   3. Others.
E. Tumors of bone marrow.
   1. Myeloma.
   2. Malignant lymphoma.
F. Bone Tumors of yet undetermined origin and incompletely characterized.
   1. Myelolipoma of the cat.
   2. Benign mesenchymoma of the medullary cavity of the cat.
   3. Multilobular tumor of the head mostly arising at suture lines of the skull and incisivomaxillary and palatomaxillary suture of the hard palate of dogs.

III. **TUMOR-LIKE LESIONS.**

A. Fibrous dysplasia.
B. Solitary bone cyst.
C. Juxtacortical bone cyst.
D. Epidermoid cyst of the phalanx.
E. Myositis ossificans.
F. Villonodular synovitis.

Note While the clinical behavior and histologic features of pleomorphic sarcomas of bone and undifferentiated high-grade sarcomas are described in man in 2013 World Health Organization Classification of Tumors of Soft Tissue and Bone (Man), Vol 120, Issue 12, June 15, 2014 pp. 1763–1774, we anticipate that these two rare tumors in animals well be reported in the next WHO Classification of Bone and Joint Tumors of Domestic Animals.

However, the authors of this textbook have examined 14 examples of eight pleomorphic sarcomas and six undifferentiated high-grade sarcomas whose radiographs and histopathology specimens were submitted to a Bone Infarct Teaching and Research Archive at Department of Pathology TAMU CVM over a period of 15 years. These submissions were from amputated long bones of dogs ranging from 5 to 10 years in age. Specimens were from 11 different large breeds of dogs of which more were from neutered male dogs than female dogs. Specimens were contributed from private practices in the USA and foreign countries by clinicians who had not received definitive radiographic and histopathology diagnoses from their original referrals. This was a collection from a study of 653 dogs having radiographic and histopathologic evidence of destructive radiographic lesions in a bone infarct. Both categories of pleomorphic sarcomas appeared to arise by neoplastic transformation of bone marrow mesenchymal stem cells that form a mixed population of hyperchromatic spindle and polygonal cells. Tumor cells formed no matrix or distinctive morphologic patterns. Those diagnosed as pleomorphic sarcoma had a mitotic count of less than 1/3 HPFs and exhibited minimal pleomorphism while unclassified high-grade sarcomas often had several mitotic figures in each HPF. These high-grade tumors were present in microcavities in necrotic cancellous bone that was undergoing active resorption by osteoclasts.

2.10 Tumors of Bone in Animals: Categories, Frequency, and Clinical Behavior

Primary bone tumors are relatively common in domestic animals and their exact frequency is still unknown. A study published in Acta Vet Scand. V. 59 in 2017, reported osteosarcoma incidence rates in dogs is 27 higher that in people. There is big variation of the related incidence between different species of animals. Dogs are the species with the highest prevalence of bone tumors among all other species of animals.

In human medicine the classification systems of bone tumors first use the specific histogenetic origin of the tumor, and categorize them into chondrogenic or cartilaginous tumors, osseous or osteogenic, fibrogenic or fibrous tissue bone tumors, fibrohistiocytic tumors, Ewing Sarcoma, hematopoietic neoplasm, osteoclastic giant cell-rich tumors, notochordal tumors, vascular tumors, myogenic, lipogenic, and epithelial tumors, tumors of undefined neoplastic nature, and undifferentiated high-grade sarcomas. By comparison veterinary medicine uses the animal WHO classification system. The animal WHO Classification first considers for a diagnosis the nature of the tumor as benign or malignant. It then considers the cell of origin as a characterizing factor in definitive diagnosis. On this basis veterinarians recognize three main categories of bone tumors: benign tumors, malignant tumors, and tumor-like lesions within which all the various and possible neoplastic conditions that can arise at bone level in domestic animals are represented (Brown et al. 1986; Carrig and Seawright 1969; Enneking 1983; Heck et al. 2008; Morgan et al. 1980; Sullivan 1960; Szymendera 1970; Wold et al. 2003; Strafuss 1987) (Table 2.1).

Table 2.1 Different general symptoms/lesions/signs generally potentially characterizing bone tumors

Pain
Swelling
Pathological fracture
Neurological signs (Symptoms)
Metastases
Infections
Hematological disorders (e.g., Alkaline phosphatase)
Incidental findings (e.g., benign tumors)

2.11 Benign Tumors of Bones

2.11.1 Osteoma

2.11.1.1 Definition

Osteoma is a relatively uncommon benign solitary nodular bone mass that arises from the cortical surface of a bone in the head often from either the buccal or lingual surface of the body of the mandible. This solitary smoothly contoured mass is covered by the fibrous layer of the periosteum that overlies a mass of cancellous bone, i.e. cancellous osteoma, or compact bone, i.e. a dense or "eburnated" osteoma. They do not invade the cortical surface or undergo neoplastic changes. These benign bone tumors are distinguished from a traumatically-induced exostosis, or other benign and malignant periosteal tumors, e.g. periosteal chondroma, parosteal osteosarcoma, periosteal osteosarcoma, etc. Bone island is osteoma-like nodular lesion located in the medullary cavity of a bone that is a residual bony mass that was not remodeled and removed during skeletal development.

2.11.1.2 Epidemiology

Osteoma is a benign nodular bone lesion that in animals more often arises from the cortical bone surface in the skull than from other bones of the skeleton. In the veterinary literature osteoma has been more frequently reported in horses and infrequently reported in other domestic animals. However, with the recent importance of dental surgery in domestic pets, osteomas are being increasingly recognized in cats and dogs.

2.11.1.3 Localization

Osteoma is a benign osteogenic nodular lesion that arises from cortical surfaces of craniofacial bones of intramembranous origin of the calvarium, sinuses, and mandible and less frequently from bone surfaces of the axial and appendicular skeleton. Comparative terminology differences between animals and man partially reflects changing nomenclature with different editions of the WHO classification systems. The most recent WHO bone tumor classification for man only incudes: (1) Benign osteogenic tumors as osteoma and osteoid osteoma and (2) Intermediate-grade locally aggressive tumor as osteoblastoma. Three malignant osteogenic tumors arising from a cortical surface are 1) Parosteal osteosarcoma, 2) Periosteal osteosarcoma, and 3) High-grade surface osteosarcoma. Parosteal osteoma was a term formerly used to describe what was considered to be a benign counterpart of the more well recognized parosteal osteosarcoma. That entity is likely now included with osteoma. In addition to the previously indicated locations in the head, osteomas are sessile nodular benign bone tumors that arise on cortical surfaces of bones of the axial and appendicular skeleton. Osteoma arising on a long bone needs to be differentiated from, among or nodular masses on a bone surface, a sessile developmental osteochondroma, periosteal chondroma {not in the WHO man but present in dogs and cats}, parosteal osteosarcoma, and a mature focal lesion of myositis ossificans attached to the bone surface.

2.11.1.4 Symptoms

In animals an osteoma is a solitary nodular lesion that protrudes from a cortical bone surface in the head or bone surfaces of skeleton of the animal. In man while a solitary lesion is the common presentation, multiple osteomas are findings in Gardner syndrome associated with familial multiple intestinal adenomatous polyposis. Osteomas are generally asymptomatic nodular lesions arising in flat bones of the skull and sinuses where by their size, they can impinge on adjacent structures, e.g. obstruct passages, press on the brain, etc. Small animals with osteomas arising from the lingual or buccal surfaces of the mandibular body are often presented to veterinary dentists when owner detects a swollen jaw or the animal has problems eating because the mass compresses the tongue. Osteomas on limbs of animals are usually nonpainful masses unless they are traumatized, interfere with a connective tissue attachment, or are palpated by the owner.

2.11.1.5 Gross Pathology

Osteoma generally has oval, round, or hemispheric smooth surface covered by a periosteum, exophytic growth, and has a contoured border covered by vascularized mesenchymal hard tissue. In cats and dogs that can arise on the outer and lingual surfaces of the mandible with oral osteomas causing difficulty in eating and interfering with the tongue. Osteomas arising from the skull can resemble nodular lesions caused by multilobular skull tumors and when arise from the sinuses may have edematous or mucinous aspects. In the rare intramedullary forms of osteoma is possible to find small islands of compact bone fused with different bone trabeculae. In horses an osteoma can also reach very large dimensions. In the human species it is described with dimensions that generally remain smaller than the dimensions of 2 cm in diameter. The color may vary from tan to white and the mass resembles the cortical bone where it merges. Sometimes well-formed subperiosteal reactive bone may be present in triangular-shaped lesions.

2.11.1.6 Cytology Features

Considering the frequent sclerosis that can be present on the periphery of these tumors, it is very difficult to obtain a good cytological preparation from a needle. In some cases, reasonable results can be obtained with imprinting cytology. Needle aspirates specimens are often very moderate cellularity with cells mixed with a background of red blood cells. The most representative cells are osteoblasts, pink-orange fibrillar matrix with several osteoblastic cells that may line the surface of the matrix fragments. Mitotic figures are absent and in cases of osteoid osteoma occasional multinucleated giant cells could be detected.

2.11.1.7 Microscopic Pathology

While many long-standing osteomas are composed of dense bone, some are cancellous and some still forming osteomas are initially cancellous before becoming dense osteomas. The contoured surface of an osteoma is covered by a periosteum that has a fibrous and osteogenic layer during formation and may only have a fibrous layer of the periosteum when mature, i.e. no longer forming. The histologic features of osteoma are represented by a mixture of lamellar and woven bone also accompanied by Haversian and Haversian-like systems (Fig. 2.1). The tumor is composed by abundant spicules and trabeculae of cancellous bone (Fig. 2.2) that should become compact bone with time. Not always osteoma consists of interconnecting trabeculae of bone and may contain fibrous component as like as in fibro osseous lesions.

Osteoblastic cells represent the prevalent cellular component and appear with a plasmacytoid plump cells morphology with oval to round nuclei. Osteoblasts that rimming the bone tissue are moderate flat and small in size. Actively growing lesions usually contain plump active osteoblasts that have abundant eosinophilic cytoplasm and nuclei polarized distant from bone-forming surface. Osteocytes have nuclei with condensed chromatin and often small and no visible nucleoli. Osteoclast and osteoclast-like giant cells may be mixed in the lesion and mitotic figures represent only rare and occasional finds. Usually, the central nidus of primary osteoma is surrounded by dense and mature sclerotic tissue (bone).

It is important to remember that the histological aspects of the lesion can also vary in relation to the time of onset/permanence/stage. In early osteoma there is a little bone production and the osteoblast proliferation is active in a stroma highly vascularized. In an intermediate stage the tumor is characterized by abundant osteoid and a smaller fibrovascular stroma. In the last mature phase, the dominant histopathological character of the lesion is represented by osseous compact trabeculae highly calcified with an irregular arrangement.

Core needle biopsy of osteoma, or "osteoid" osteoma should be characterized by a meshwork of new bone, osteoid, osteoblast, and osteoclast-like giant cells that are almost present.

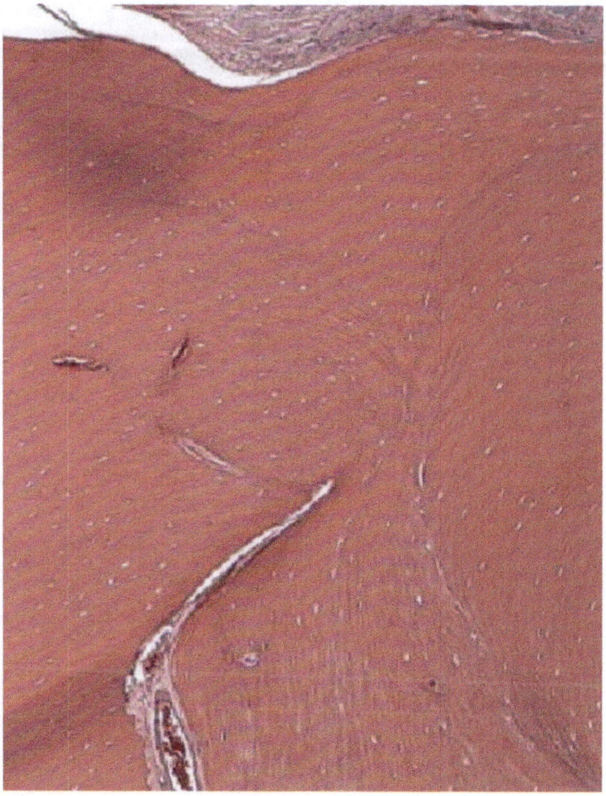

Fig. 2.1 Dense or eburnated osteoma from mandible of cat (Courtesy of Dr. Roy R. Pool, DVM, PhD, Emeritus Professor of Pathology UCVD and Texas A&M—USA)

2.11.1.8 Differential Diagnosis

Bone-forming lesions

- Osteochondroma: has a cartilage cap.
- Parosteal osteosarcoma: may contain malignancy characters and spindle cells components but should lie superficial to the osteogenic layer of the periosteum and not be attached to the cortical surface like an osteoma.
- Juxtacortical myositis ossificans: composed of cancellous bone containing cartilaginous and fibrous elements and should be entrapped in muscle fascia.

2.11.1.9 Diagnostic and Interpretative References List (Diagnostic Pearls)

- Composed of well-formed and well-differentiated bone without malignant cellular atypia (osteoma versus osteosarcoma);
- The intact cortex and absence of cartilage and cartilaginous cap excludes osteochondroma;
- The presence of low number of cells allows and helps to exclude a form of myositis ossificans.

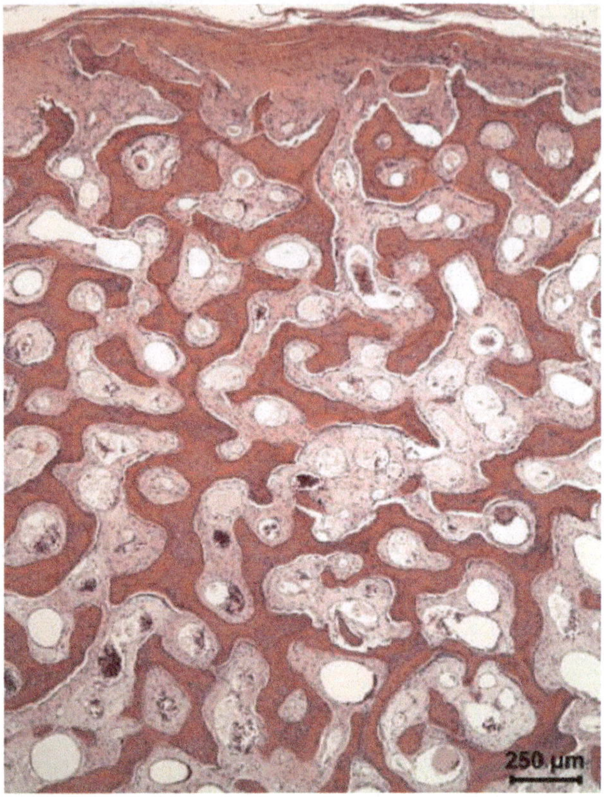

Fig. 2.2 Cancellous osteoma from mandible of a cat (Courtesy of Dr. Roy R. Pool, DVM, PhD, Emeritus Professor of Pathology UCVD and Texas A&M—USA)

Ancillary Techniques.

Immunohistochemistry and/or molecular diagnostics investigations have not determinant role in the work-up of osteoma.

Osteoma

Definition

Benign bone-forming tumor, composed of dense lamellar well-differentiated mature bone.

Clinical features

Small and solitary and frequently in craniofacial skeleton.

Usually asymptomatic.

Radiographic features

Small dimension and radio dense mass.

Well marginated and well-formed periosteal reaction.

No involvement of underlying cortex.

Dome shape with large attachment to the bone cortical surface.

(continued)

MR Findings: *lesion does not enhance with contrast.*

CT Findings: *round to oval shape well delineated bone surface mass with cortical appearance.*

Macroscopic features

Round, hard, tan-white, and small in diameter.

Microscopic pathology

Mainly composed of lamellar bone with less quantity of woven bone.

The main architecture is cortical-type and less of trabecular type.

Intralesional cells (osteoblasts, osteocytes, osteoclasts) are usually inconspicuous.

2.11.2 Osteoblastoma Not Classified in WHO of Animals

2.11.2.1 Definition

Osteoblastoma is a small benign bone-forming tumor that has been described as occurring in three cats, two dogs, mice, a pony, and a hedgehog. Lesions in the distal humerus and spine in two cats reached up to 8 mm in diameter (Kirk et al. 2019; Giebels et al. 2020). In man this tumor may reach more than two centimeters in size.

In man osteoblastoma is very similar in histologic appearance to osteoid osteoma where it is composed of woven bone spicules always bordered by osteoblasts. Osteoblastoma and osteoid osteoma are very similar and their differentiation is based on size of the mass, location of the tumor, clinical features, and radiographical and image findings. Because of so few osteoblastomas and osteoid osteomas have been documented in animals their skeletal location and radiographic features have not been systematically characterized and aggressive variants in animals may have been interpreted as low-grade osteosarcoma.

"Aggressive Osteoblastoma" (only described in man) has macroscopic dimensions greater than 4 cm. in diameter and is characterized by destructive growth and high rate of recurrence without metastatic potential. Histologically the most significant differences of aggressive osteoblastoma from osteoblastoma are the presence of atypical round cells with abundant eosinophilic cytoplasm with eccentric large oval nucleus with prominent nucleolus called as Epithelioid osteoblasts.

2.11.2.2 Epidemiology

Osteoblastoma is not classified in WHO classification of bone and joint tumor of domestic animals but rare cases in animals were described by Kirk et al. in 2020 in a cat, by Goedegebuure et al. in 1983 in a pony, and by Kim et al. in 2017 in a cat. This rare lesion in animals was described as a comparison for this lesion in man. In humans osteoblastoma and other osteoid tumors represent approximately a range of

3%–12% of all human benign bone tumors, with most cases diagnosed in the first three decades of life.

2.11.2.3 Localization

Osteoblastoma arising within the cancellous bone and commonly affects long bones as femur and tibia and then with less frequency vertebral column (transverse process or neural arch), facial bones, ribs, and digit.

In human species and in the tubular long bones the primary site of onset is represented by metaphysis or diaphysis. In humans there are also very rare descriptions of aggressive multicentric osteoblastoma.

2.11.2.4 Symptoms

The general clinical signs recognized for this tumor are a presentation of pain, swelling, low decreased motion, neurological symptoms when osteoblastoma is primarily localized in spine (torpor, paraparesis, paraplegia, etc.).

2.11.2.5 Gross Pathology

In man osteoblastoma appears as a solitary lytic, round, or oval, tan-white, dark-red, well-circumscribed, gritty mass, frequently covered or circumscribed by periosteal reactive bone. When appear as a "large" tumor it can produce also aneurysmal bone cyst (10% of cases in human species). The size range of the tumors in man is 2–20 cm. The outer edge ofosteoblastoma usually is well defined with no permeative behavior as like as in the different forms of osteosarcomas. A frequent common character of this tumor is represented by hyper vascularity especially when tumor arise from cancellous bone and only in rare cases is possible to detect hematic cavities typical of aneurysmal bone cyst. Often osteoblastoma has a "pushing" border against the cortical surface and trabecular bone of the marrow.

2.11.2.6 Cytology Features

Osteoblastoma and other osteoid osteomas practically are never diagnosed cytologically but we can report some features described for human species. High cellular needle aspiration from osteoblastoma usually contains osteoblastic cells, with eosinophilic cytoplasm (epithelioid osteoblasts), with eccentric nuclei. Some of these cells contain a well-demarcated halo referring to prominent Golgi as like as frequently detected in activated osteoblastic cells. Mixed with these cells, scattered multinucleated giant cells are frequently detected. Mitotic figures are rare.

2.11.2.7 Microscopic Pathology

Osteoblastoma has similar histological features of osteoid osteoma. This tumor is composed of neoplastic woven bone spicules and thin trabeculae deposited in irregular interconnecting or sheet-like or organoid pattern. Large osteoblastic cells producing osteoid and woven bone. Osteoblasts rimmed the trabeculae and also osteoclasts should be detected. Mitotic figures are rare. Osteoblasts have round or ovoid shape with moderate amount of eosinophilic, sometimes purple, cytoplasm, and eccentric nuclei with fine chromatin. Occasionally cells with bizarre hyperchromatic nuclei may be present. Necrosis normally is absent or rarely focal. In a rare cartilaginous variant contain well-formed cartilage (hyaline). A very important finding of osteoblastoma that differentiates it from osteosarcoma is the absence of permeative pattern and that the interface between osteoblastoma and surrounding bone is very sharp. Just in rare cases osteoblastoma has been described with multifocal (or multinodal) growth (Fig. 2.3).

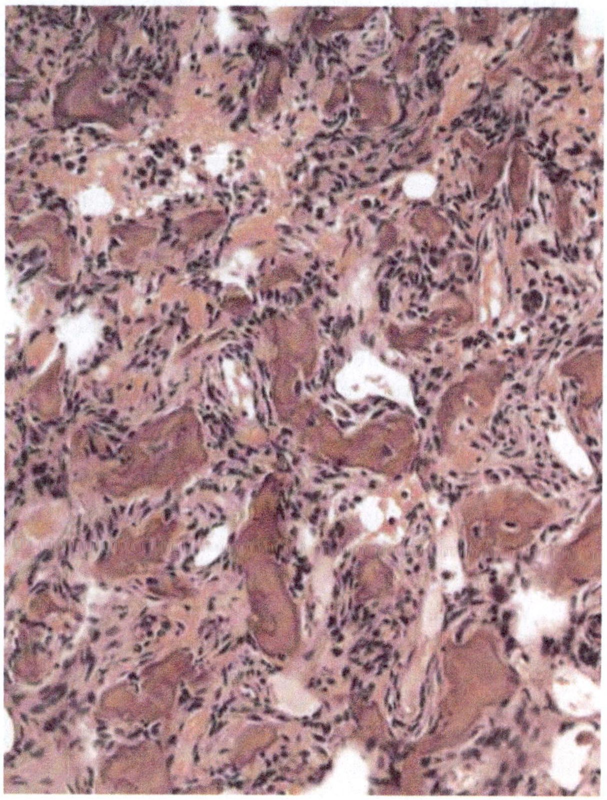

Fig. 2.3 Osteoblastoma of distal humerus of a dog (Courtesy of Dr. Roy R. Pool, DVM, PhD Emeritus Professor of Pathology UCVD and Texas A&M—USA)

2.11.2.8 Differential Diagnosis

- Osteoid Osteoma (in man): < 2 cm.;
- Aneurysmal bone cyst: no irregular trabeculae of woven bone;
- Osteoblastoma-like osteosarcoma: infiltrative pattern.

2.11.2.9 Diagnostic and Interpretative References List (Diagnostic Pearls)

Osteoblastoma is well demarcated with neoplastic bone always rimmed by osteoblast cells.

Ancillary Techniques.

Immunohistochemistry and/or molecular diagnostics investigations have not determinant role in the work-up of osteoblastoma.

Osteoblastoma

Definition

Benign bone-forming tumor composed of neoplastic woven bone spicules and thin trabeculae deposited in irregular interconnecting or sheet-like or organoid pattern.

Clinical features

Small and solitary commonly arises in tubular bones and spine > of 2 cm in diameter.

Characterized by pain, swelling, when localized to spine with neurologic symptoms.

Good prognosis with local recurrence in human species of about 20%.

Radiographic features

Exophytic well-defined lytic mass with frequent sclerotic margins, peripheral halo surrounds area of mineralization.

Common presence of periosteal reactive bone.

MR Findings: *well-delineated mass with mineralized areas with void signal (dark). Tumor and edema better assessable with contrast.*

CT Findings: *Exophytic lytic and sclerotic mass surrounded by reactive bone.*

Macroscopic features

Dimension ranges from 2 to 20 cm (usually 2–5 cm) in size and radio dense well-circumscribed mass, gritty, tan-white, and dark-red.

No involvement of underlying cortex.

Dome shape with large attachment to the bone cortical surface.

Round, hard, tan-white, and small in diameter.

Microscopic pathology

(continued)

Well demarcated from adjacent reactive pre-existing bone tissue, composed of neoplastic woven bone spicules and thin trabeculae deposited in irregular interconnecting or sheet-like or organoid pattern.

Presence of epithelioid osteoblasts (75%) in the aggressive form of osteoblastoma.

Vascular connective tissue fills intertrabecular space.

Variants may contain: epithelioid osteoblasts, cartilage, aneurysmal bone cyst-like changes.

2.11.3 Myxoma of the Jaw (Odontogenic Myxoma)

2.11.3.1 Definition

Myxoma of the jaw in veterinary medicine is defined as an uncommon bone tumor composed of stellate cells immersed in a gelatinous matrix that involve bones of the jaws of horses, cattle, and less commonly in dogs. Because most of these jaw tumors arise in a disturbance of the odontogenic apparatus, most of these tumors are classified as odontogenic myxoma. Myxoma involving other bones of the axial and appendicular skeleton and joints are very rare except for the synovial myxoma considered the second most common tumor of joints for frequency. Myxoma is also described in the articular process of vertebrae.

2.11.3.2 Epidemiology and Localization

Due to the extreme rarity of this tumor, there are no significant epidemiological data that allow to express significant and valid epidemiological values, except for those of the primary localization of the tumor on the jaw of large animals.

2.11.3.3 Symptoms

The general clinical signs recognized for this tumor are a presentation of pain, swelling, difficulty in eating, very slow growth in the benign form, and more rapidly invasive in the malignant form.

2.11.3.4 Gross Pathology

Myxoma of the jaw may be lobulated, not capsulated usually, and destructive lesion of mandible may be present with osteolytic lesions associated with thin bony septa.

2.11.3.5 Cytology Features

Fine-needle aspiration (FNA) of myxoma of the jaw stained with Romanowsky method generally revealed a thin film of granular or mucoid magenta material where can be present rare single and aggregates of stellate cells, sometimes spindle-shaped fibroblasts with small dark uniform dark nuclei with nuclear-cytoplasm ratio is high and inconspicuous nucleoli. The stellate cells component is characterized by long cellular and cytoplasmic processes. No mitotic figures are detected usually.

2.11.3.6 Microscopic Pathology

The histopathological features of this tumor are characterized by the presence of neoplastic stellate to spindle-shaped cells in an abundant gelatinous basophilic (myxoid) matrix that is a prominent feature in odontogenic myxomas arising in the jaw of large animals. Myxoid matrix is composed of collagen and reticular fibers embedded in a gelatinous matrix that includes proteoglycans. The multilocular pattern results from a combination of slow growth and a mechanical effort by the jaw to contain the mass by the formation of incomplete septa. Tumor margin may be either unencapsulated or partially encapsulated. Rapid invasiveness growth, cellularity, cellular atypia, and the presence of atypical mitoses distinguish myxoma from the myxosarcoma.

2.11.3.7 Differential Diagnosis

- Non-odontogenic tumors elaborating myxoid matrix (fibrosarcoma and odontogenic fibroma).

2.11.3.8 Ancillary Techniques

Immunohistochemistry and/or molecular diagnostics investigations have not determinant role in the work-up of myxoma/odontogenic myxoma.

High cellularity and mitotic count combined with larger cells and cellular atypia distinguish odontogenic myxosarcoma from myxoma. Odontogenic epithelium may be present but is not required to establish the diagnosis. Microscopic differential diagnoses include non-odontogenic tumors that may elaborate some myxoid matrix such as fibrosarcoma, which will lack close association with teeth. A peripheral odontogenic fibroma may also appear histologically similar, but will lack myxoid character of the extracellular matrix (Fig. 2.4).

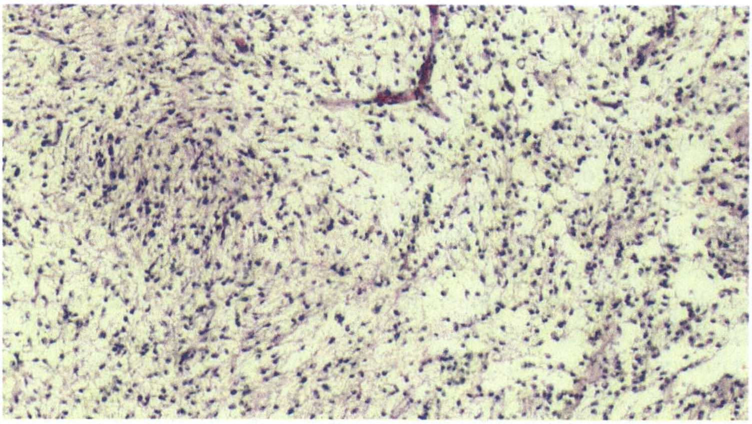

Fig. 2.4 Myxomatous pattern typical of myxomas with neoplastic stellate to spindle-shaped cells in an abundant gelatinous basophilic (myxoid) matrix. Synovial myxoma from a dog (Dog, H&E, from author collection)

Myxoma of the Jaw

Definition

Benign bone-forming tumor, an uncommon bone tumor composed of stellate cells immersed in a gelatinous matrix that involve bones of the jaws.

Clinical features

Pain, swelling, difficulty in eating, very slow growth in the benign form, more rapidly invasive in the malignant form.

Macroscopic features

May be lobulated, not capsulated usually, and destructive lesion of mandible may be present with osteolytic lesions associated with thin bony septa.

Microscopic pathology

Neoplastic stellate to spindle-shape cells in an abundant gelatinous basophilic (myxoid) matrix that is a prominent feature in odontogenic myxomas arising in the jaw of large animals.

2.11.4 Osteochondroma

2.11.4.1 Definition and Pathogenesis

It is classified as a benign tumor of the surface of the bone cartilage-capped and formed by an underlying proliferative outgrowth of endochondral ossification. Osteochondroma originates frequently in the regions of metaphysis or apophysis and occurs essentially in two main forms: as a solitary tumor and as a tumor with multiple lesions (also called multiple hereditary osteochondromatosis) that represent an autosomal dominant genetic disorder that is related to inherited autosomal dominant disease described in horses, pigs, and dogs. There is a great and old debate about the origin of this lesion: someone consider osteochondroma as a pure tumor, someone consider it as an abnormal proliferation of cartilage and bone due to a kind of dysplasia of the growth plates. There are interesting data reported in humans where osteochondroma was considered to originate from a displacement or redirection of growth plate or was believed as a defect of periosteal cuff surrounding the surface of the growth plate. An osteochondroma never arises on cortical surface of a bone once it has formed. These data are also supported by the fact that osteochondroma can be experimentally reproduced by manipulation of growth plates and in particular with transplanting small fragments of physeal growth plates in the periosteal surface. Other studies in human species revealed a significative mutation in neoplastic chondrocytes of osteochondroma characterized by mutations of EXT1 and EXT2 genes on chromosome 8q24 and 11p11–12. These genes are strongly involved to encode the co-polymerase actively involved in heparan sulfate biosynthesis also if the exact alteration of the proteoglycan synthesis and related effects in abnormal cartilage proliferation is still unknown. The mutation of EXT1 and EXT2 genes strongly support also the data that chondrocytes derived from a kind of monoclonal proliferation of cells that resides in cartilage cap. In some cases of human osteochondroma were also described an involvement of EXT3 on chromosome 19p. It is always necessary to consider other differential diagnosis in case of suspect of osteochondroma to don't confuse other non-neoplastic lesions with this. The frequent ones are represented by reaction to local trauma that may induce bone and cartilaginous proliferation, callus from some kinds of fractures, anomalous cartilaginous proliferations of soft tissues that undergo endochondral ossification as like as frequently appear in synovium of joints with synovial osteochondromatosis.

2.11.4.2 Epidemiology and Localization

Osteochondroma is quite common in domestic animals especially in canine species where it is detected sometimes casually during clinical and radiographic investigations. In dogs and horses osteochondroma affected young animals that have active growth plates. In domestic animals there are no data about sex or breed predilection. Reports in humans indicate that osteochondroma has an incidence of being the most common primary tumor in that species, accounting with 36–50% of all benign bone tumors and 8.5–15% of all primary bone tumors. Osteochondroma lesions in man have been initially diagnosed in a wide range of ages from 8 to 77 years with an average age of initial recognition of 21 years. In that species males have a higher incidence of osteochondroma of about 2:1 with respect to females. There is also a description in literature about a case of osteochondroma in a pig with specific localization to the ribs (De Brot et al. 2013).

Typical site of occurrence of an osteochondroma is one or more vertebrae or ribs.

Ends of long bones of the appendicular skeleton are also common sites but these cartilage-capped exostoses may develop on the cortical surface of any bone formed by endochondral ossification. For example, they can arise on the surface of a flat bone of the scapula, an endochondral bone but not on the flat bones of the skull that arise by intramembranous ossification.

2.11.4.3 Symptoms

Sometimes many cases of osteochondroma are asymptomatic and should be detected incidentally during imaging during a clinical examination. The size, location, and clinical signs may play an important role in clinical management.

The common frequent symptoms are generally represented by pain, lameness, and in severe case also by paralysis if the mass causes compression of nerves and adjacent organs or tissues. Clinical signs have been described in dogs with vertebral osteochondroma causing neurological problems for spinal cord compression (Doige 1987; Caporn and Read 1996; Bhatti et al. 2001).

Some deformities as like as shortness of limb or angular limb are described in dogs and humans affected by osteochondromatosis (Mozos et al. 2002).

Malignant transformation of osteochondroma should be considered especially in cases with fast growth of the mass and seem that these cases occur frequently in older dogs. Keith G. Thompson and Keren Dittmer in Meuten Tumors in Domestic Animals report a case in young dogs with occasional localization in trachea without relationship between the tumor and related skeletal origin.

Osteochondroma exhibits slow growth that ceases prior to skeletal maturity. In human species malignant degeneration of a solitary osteochondroma is considered uncommon with a ratio of 0.4–2.0%.

Gross Pathology Macroscopic appearance of a solitary osteochondroma is characterized by smoothly contoured exostosis having a sessile or pedunculated stalk that arises continuously from the cortical surface with radiological feature reflecting bone density relative to the stage of development and age of this periosteal nodular lesion. The typical gross appearance of osteochondroma consists of a thin cap of cartilage (from 0.1 to 3.0 cm.). Periosteum surrounded the stalk of cancellous bone consisting of mature well mineralized bone mixed with spongiosa depending upon the stage of lesion development. Radiographic findings reveal that the mass is located in the most superficial areas of the bone sometimes in direct continuity with the underlying cortex and marrow cavity. Usually, the external surface of the cartilaginous cap is partially mineralized at its junction with it supporting bony stalk. The junction of the deep central part of cartilage cap that is in contact with the developing bony stalk should be composed of cancellous bone due to since growth of the lesion is by endochondral ossification.

Cytology Features and Histopathology Typical lesion of osteochondroma in its active stage is represented by an irregular cartilage cap with similar features of normal growth plates. In the end this kind of lesions, especially in human species are considered as a kind of hamartoma or exostosis that differentiates and grows by endochondral ossification. The external cartilage cap may exhibit lobular structure as may arise in any cartilaginous nodular lesions. The cartilage caps of a few osteochondromas may contain numerous large irregular-sized chondrocytes that are not well organized and distributed. Some features of osteochondroma show chondrocytes hypertrophy, increased cellularity, and plumpness of nuclei. At high magnification it is possible to recognize in the internal parts dense and irregular patterns of the cancellous bone often associated with osteosclerosis of trabeculae. In the active and proliferating form of osteochondroma it is also possible to find focal concentrations of osteocyte.

With higher magnification sometimes is possible to recognize signs of malignant cell transformation. The external surface of the mass usually is covered by periosteum rich in fibers, surrounding the cap of hyaline cartilage composed by a large amount of basophilic matrix, containing chondrocytes often densely packed and also arranged frequently in columnar structures that frequently taken a finger-like processes in adjoining bone tissue where often is possible to find also internal islets of cartilage in a typical pattern of osteochondroma. This histological structure makes the diagnosis of osteochondroma very easy especially with the support of radiological findings.

Differential Diagnosis There are several differential diagnoses to consider for osteochondroma. Periosteal chondromas may arise on a bone surface (periosteal chondroma). This tumor arises beneath the periosteum on the intact cortical surface of a bone. It is composed of disorganized lobules of cartilage without radial orientation of the chondrocytes in the deepest region where endochondral ossification takes place in an osteochondroma but which is not present in a periosteal chondroma. Osteosarcoma of the surface (parosteal osteosarcoma) arises from the external surface of the cortex. Non-continuity between medullary cavity and the mass is usually detected. A peculiarity of cartilaginous cap of parosteal osteosarcoma is represented by a more marked disorganization and moreover the tumor contains a lot of spindle-shaped malignant cells in the neoplastic bone. In differential diagnosis should also consider the rare periosteal myositis ossificans that involves the bone surface but without formation of cap of hyaline cartilage and where cortex remains intact.

Diagnostic and Interpretative References List (Diagnostic Pearls) The most significant findings are represented by the presence of cartilaginous cap with endochondral ossification and identical signs of normal growth plate. Some authors suggest that if cartilage cap is greater than 1 cm in thickness, the tumor has high chances of acquiring malignancy and turning into chondrosarcoma, but not all community of scientists accept this sentence and consider that cartilage tissue should acquire atypical lobular infiltrative growth pattern with increased cellularity and clear characters of cellular malignancy (Fig. 2.5).

Fig. 2.5 Periosteal chondroma forming a mass on the surface of a leg bone of a cat (Courtesy of Dr. Roy R. Pool, DVM, PhD Emeritus Professor of Pathology UCVD and Texas A&M—USA)

2.11.5 Feline Osteochondromatosis (Syn. Multiple Cartilaginous Exostosis (MCE))

As reported in WHO Histological Classification of Bone and Joint Tumors of domestic animals: "*because of the more serious consequences of lesions in cats, it is important to note that osteochondromas in dogs and horses never occurs on bones of intramembranous origin, such as in the skull. In cats, however, the tumors arise after skeletal maturity, have a random distribution including the skull, demonstrate progressive growth like a tumor, and have viral particles resembling feline leukemia virus (FELV) and feline sarcoma virus (FSV) in the proliferating cells located at the surface of the lesions.*" This initial description of this disorder provided illustrations of budding c-type retrovirus having ultrastructural features of FELV and FSV. Viral particles were present in proliferating chondrocytes located in the fibrous periosteum of a cat having random skeletal distribution of the exostotic lesions including the intramembranous flat bones of the cranium (Pool and Carrig 1972). In that initial cat and for other cats with MCE with viral particles in their lesions examined by those authors, there was no evidence of lymphosarcoma or sarcoma at necropsy of those cats. This suggests that this virus while having ultrastructural features of FELV and FSV is a separate viral agent.

The involvement of the viruses in the etiopathogenesis of feline osteochondromatosis has not been experimentally reproduced. However, the localization of the lesions in the periosteum and the presumptive activation and proliferation of resident mesenchymal stem cells in the periosteum, and the random distribution of the lesions are important factors supporting the hypothesis of viral etiology in cats.

There are few cases of serological negative Feline Leukemia Virus (FeLV) in described cases of feline ostechondromatosis. The cats with osteochondromatosis have a range of age from 16 months to 8 years without any sex or breed predisposition. As with lesions of developmental osteochondroma/osteochondromatosis in foals and pups, the initial lesions of MCE can also be initially asymptomatic until their presence is manifest due to progressive growth. Owners recognize the nodules during palpation or if they interfere with function. The diffuse random skeletal distribution separates the feline MCE disorder from developmental osteochondroma/osteochondromatosis of pups and foals.

The lesions of feline osteochondromatosis are characterized by diffuse, dense, bone masses made by one or multiple nodules with large and sessile base and formed by cancellous bone. Sometimes the nodules are coated also by a layer of thin cartilage. The nodular mass arises from the external surface of the bone and form an exophytic tumor. The histologic features of a developmental osteochondroma differ from that of an exostotic MCE nodule. The developmental osteochondroma retains some basal features that mimic the linear arrangement of a growth plate from which the development osteochondroma arose resulting from a disturbance of the perichondrial ring. By contrast the more orderly patterns of endochondral ossification at the base of the lesion are absent less distinctive in lesions of MCE since the cartilage nodule of MCE arose from a proliferation of chondrocytes that formed as a result of a neoplastic change in a mesenchymal stem cell present in the fibrous periosteum. The cartilage of the MCE nodule is more cellular and less orderly and its mineralizing margin undergoes endochondral ossification. Because of the mesenchymal stem cell origin of the MCE, that stem cell can give rise to cellular clones of cells that form different histotypes represented by adipocytes, connective fibrous tissue, and hematopoietic tissue.

Summary: in the feline species the lesions of multiple cartilaginous exostosis (syn. Osteochondromatosis or MCE) have unique etiopathogenetic characteristics, due to a disparity in their etiopathogenesis yet undergo endochondral ossification that forms a bony base for both types of exostotic bone lesions.

Unlike what happens in a developmental osteochondroma where the lesions can cease enlargement upon reaching skeletal maturity, feline osteochondromatosis exhibits progressive growth as happens in neoplasia. A description of malignant transformation of lesions of MCE to osteosarcoma has also been reported in a cat (Doige 1987), (Figs. 2.6, 2.7, and 2.8).

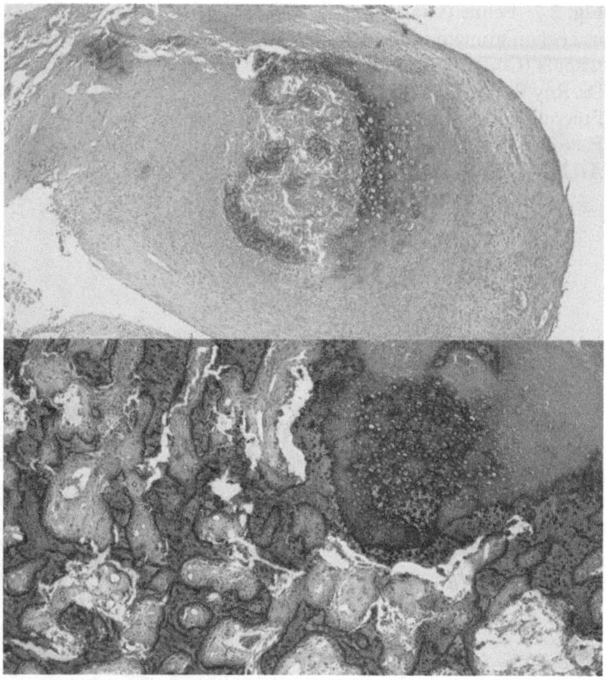

Fig. 2.6 Feline osteochondromatosis, Cat, European, 1 year old (H&E). Diffuse, dense, bone masses made by one or multiple nodules with large and sessile base and formed by cancellous bone. The top picture shows a nodule coated also by a layer of cartilage (from author collection)

2.11.6 Chondroma of Bones of the Skeletal Surface

2.11.6.1 Chondroma of Bone and Periosteal Chondroma

Explanation for this categorization: In veterinary medicine where diagnosis of skeletal tumors is often resulted from examination of limited biopsy specimens without accompanying radiographs for comparison, these benign superficial cartilage masses have been diagnosed as "chondroma" based upon their benign histologic appearance of the available biopsy tissue. While many of these specimens were likely from periosteal chondromas, there are an undetermined number of benign cartilage tumors referred to here as "Chondromas of Bone" that appear to have arisen within the cortical compacta, but not in the medullary cavity. These "Chondromas of Bone" have subsequently extended peripherally into and have elevated the periosteum vs "Periosteal Chondromas of Bone" that have clearly arisen in the periosteum and have also extended from the bone surface.

Definition and Pathogenesis Chondromas are defined as a relatively uncommon benign cartilaginous tumors thought to originate from neoplastic transformation of mesenchymal stem cells in cartilage. These benign tumors are composed mostly of lobules of well-differentiated mature hyaline cartilage usually demarcated by a periosteum that forms a capsule of well-formed reactive bone. Chondroma occurs only in bones preformed in cartilage, i.e., by endochondral ossification.

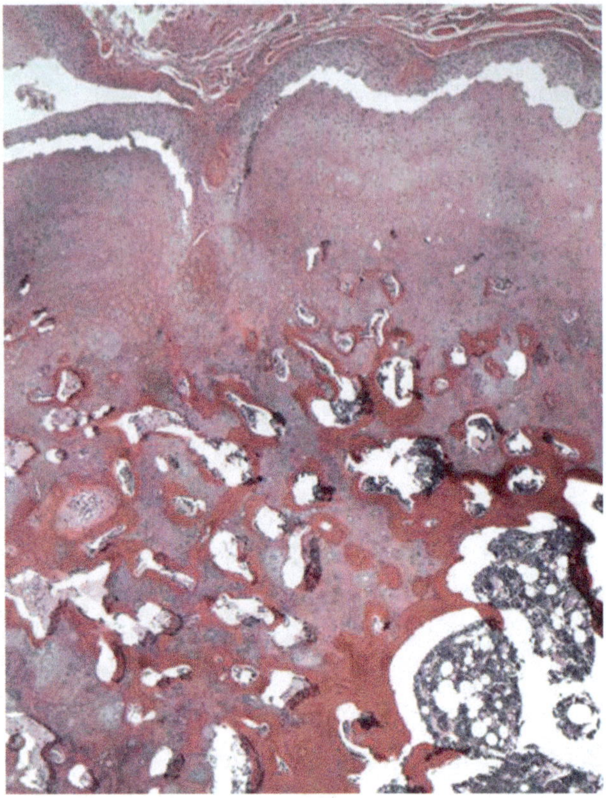

Fig. 2.7 Feline (viral) osteochondromatosis from scapula (Courtesy of Dr. Roy R. Pool, DVM, PhD Emeritus Professor of Pathology UCVD and Texas A&M—USA)

1- "Chondromas of bone" most frequently arise on surface of cortex, i.e. "Periosteal chondromas" but may also arise in connective tissue adjacent to bone or other soft tissues. Those that arise in the medullary cavity of a bone are called "enchondromas." Some classifications of these extramedullary benign cartilage tumors include: chondroma, periosteal chondroma as benign tumor examples, and periosteal chondrosarcoma as a malignant variant. Parosteal osteosarcoma which may contain cartilaginous tissue as a periosteal tumor of intermediate-grade of malignancy.

"Chondromas of soft tissue" include those that develop in sinuses and in several soft tissue sites not related to the skeleton.

Epidemiology and Localization Chondroma arises from flat bones, e.g., of the scapula, and ribs more than from long bones. Chondromas are solitary bone lesion and rarely as multiple bone lesions. Chondroma can occur in metaphysis, epiphyses, carpal and tarsal bones. In human species chondroma represents the most frequent bone tumor of the hands where chondroma has an incidence of less of 1% of all bone tumors. In Veterinary Medicine chondroma is a rare tumor described mostly in old

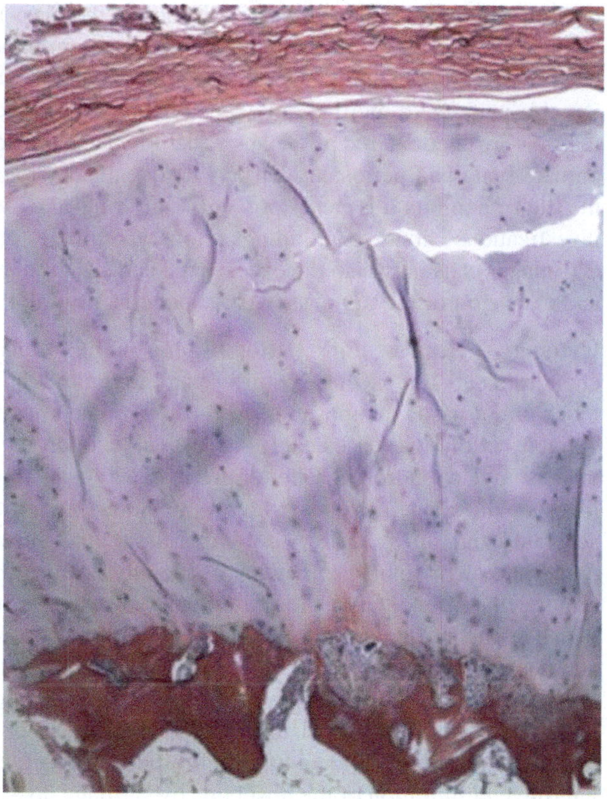

Fig. 2.8 Developmental osteochondroma from the distal end of the radius of a 6-year-old horse (Courtesy of Dr. Roy R. Pool, DVM, PhD, Emeritus Professor of Pathology UCVD and Texas A&M—USA)

dogs and sheep without any sex or breed preference. Surfaces of tubular bones represent the most characteristic sites for chondromas.

Symptoms Usually chondroma is painless for its typical slow and limited growth, for its low degree of surface reaction and for the low grade of vascularization. Pain can occur if tumor growth interferes with tendons or tendon insertions by tumor growth or causes stress or pathologic fractures. Even in rare cases chondroma involving vertebrae may cause a compression of the spinal cord clinical signs. Pathologic fracture can occur especially if small bone segments are involved.

Gross Pathology presentation of chondroma is represented by a firm swelling caused by a mass of gray-white cartilage within a bone that may cause cortical expansion while the surface of the tumor is usually covered by the periosteum. The dimension of chondromas depends somewhat on the size of involved bones, e.g. usually smaller lesions in ribs and bones of the feet, big when arise while larger in long bones of the appendicular skeleton. In an older chondroma internal calcification may also be present. The cutting surface can show variable aspects which include a compact and homogeneous whitish surface versus mottled or multinodular patterns. The neoplastic mass tends to have a whitish color in most cases also including color variables with variegations of the mass tending to bluish due to the abundant presence of cartilaginous matrix. Radiographic findings commonly demonstrate superficial radiodensity bordering a central radiolucent lesion. However, variable patterns of mottled radiodensity and radiolucency with and without defined margins are encountered as well as variability as to erosion or density of the adjacent cortical bone and periosteal density. Frequently, radiographs will demonstrate that the tumor exhibits a typical "pop-corn" or "ring-like" well-defined pattern. In MR investigations chondroma findings are very similar to other benign cartilage tumor. CT images can show also the exact shape and size of the mass. CT permits detection and extension of the tumor as well as patterns of tumor calcification.

Cytology Features and Histopathology The typical histological features of chondroma are represented by diffuse and irregular lobules of disorganized cartilaginous tumor tissue composed of well-differentiated mature hyaline cartilage with of variable cellularity. Periosteal (juxtacortical) chondroma usually contains more cells than central chondroma. Chondroma has a typical lobular formation with single lobes at different grade of maturation. Sometimes the cartilaginous lobules are separated and surrounded by a fibrous tissue and fibrous stroma. Tumor cells are ovoid and have small dense nuclei generally uniform in size and rare binucleated tumor cells. Degenerative areas of interstitial tumor matrix may have myxoid tissue which attenuates the intensity of the basophilic matrix. Focal areas of mineralization and necrosis may be present in some chondromas. In human pathology Mirra J. M et al. in 1985 described as sign of benignity of chondroma. Unlike what happens in chondrosarcoma, cartilage islands demonstrate slow or arrested growth by finding tumor cells residing inside a thin shell of mature reactive lamellar bone. That report interpreted infiltration of the host bone by cartilage does not indicate malignancy but presumably by advancement of the semi-soft interstitial cartilage matrix during growth and expansion of the tumor mass into adjacent soft tissue spaces.

Differential Diagnosis In the absence of representative biopsy material, the diagnosis of chondroma may only be presumptive when based up the location, clinical findings of chronicity and slow growth and radiographic features of benignity based upon orthogonal radiographic views. The differential diagnosis for chondroma is chondrosarcoma. Histopathological differentiation between the two may require multiple tissue specimens from different locations in the tumor and examination for previously described cellular features of malignancy, e.g. cellular atypia and spindle cell differentiation not present in chondroma. Radiographic features and clinical findings should be most helpful in distinguishing "chondroma of bone" from periosteal tumors having some cartilage components in them such as an "osteochondroma," "periosteal chondroma," "periosteal chondrosarcoma" and "parosteal osteosarcoma," or a "soft tissue chondroma" arising in extraosseous soft tissues of the limbs, e.g. ossifying myositis lesions that contain cartilage or chondromas arising in other organs and soft tissues of the body where chondromas arise (Figs. 2.9 and 2.10).

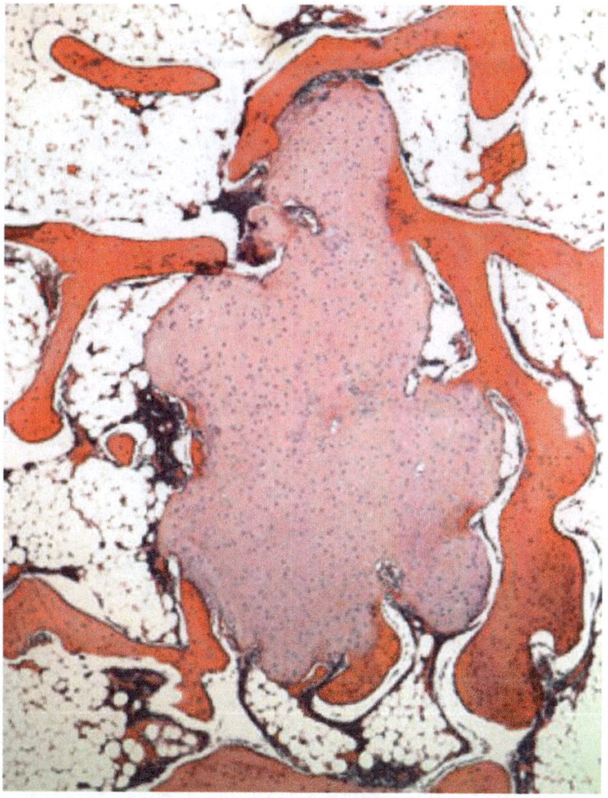

Fig. 2.9 Enchondroma in the medullary cavity of the femur of a 1.5-year-old dog was an incidental finding and part of the cartilage model that was not remodeled and replaced by cancellous bone (Courtesy of Dr. Roy R. Pool, DVM, PhD, Emeritus Professor of Pathology UCVD and Texas A&M—USA)

Chondroma

Definition

A rare benign round or oval cartilage tumor of mature cartilage usually encapsulated, smooth, or nodular.

Clinical features

Clinical issues of a swelling and exophytic mass, pain, and prevalence of involvement of tubular bones (surface generally). Uncommon recurrences.

Image Findings

General appearance is characterized by a radiolucency and calcifications, clear margins, and reactive periosteal bone.

Macroscopic features

Shiny mass with gray-white cartilage, sometimes mucinous and with multilobular pattern in cut surface.

Microscopic pathology

Hyaline cartilage represents the tumoral matrix mixed with not abundant presence of neoplastic cells and chondrocytes. Not infiltrative pattern in the underlying bone.

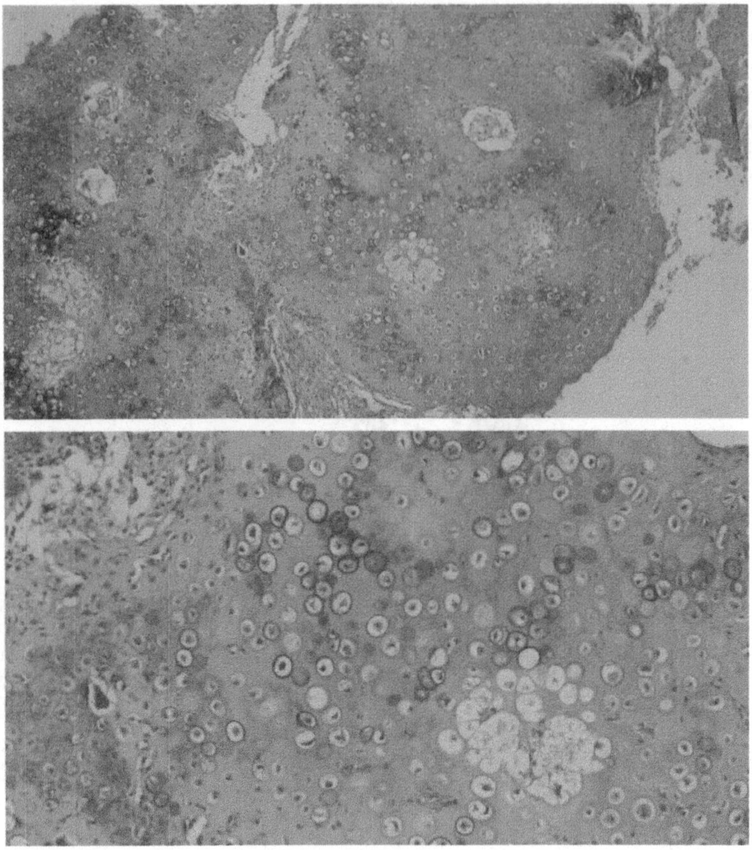

Fig. 2.10 Chondroma or chondrosarcoma? dog, H&E, 4x and 10 x magnification. From the Author: *Histological features of chondroma and chondrosarcoma have a broad range of aspects relating to the grade of differentiation and the cellular alterations and malignancy index that can vary from low-grade lesions (like in chondroma) to severe high grade more like to chondrosarcomas. Probably chondrosarcoma represents the tumor where differentiation it from the related benign form as chondroma… Progressive transformation to malignancy and the same tumor may show contemporarily different pattern of cellular modification. Small core biopsies are not always representative also in well-differentiated lesions. Clinical and imaging findings may be very helpful to evaluate macroscopic characters of the lesions and also histological investigations that give morphological and staging of the tumor* (from author collection)

2.11.7 Hemangioma and Lymphangioma

Definition and Pathogenesis Hemangioma is a rare well-differentiated benign primary bone tumor of the blood vessel origin that is lined by endothelial cells. Lymphangiomas arising in bone while also lined by endothelial cells are derived from precursor cells of lymphatic endothelial cell that line veins. Both can give rise to irregular, endothelial-lined, blood-filled structures having tubular and cavitary features (Butch and Caron 2005).

Epidemiology and Localization Described as very rare tumor in dogs, cats, and horses. In literature there are no sufficient data to hypothesize epidemiological incidences relating to sex, species, breed, and age. In human species hemangioma is one of the most common primary benign tumors of bone, with an incidence of approximately 10% of asymptomatic vertebral body's cases. In humans clinically significant hemangiomas represent less of 1% of primary bone tumors, with prevalence in women and adulthood. In animals the skeletal apparatus unlike the human species represents the sites of greatest onset where most commonly hemangioma arise in vertebral bodies (thoracic and lumbar regions) and in a lower percentage in craniofacial bones (skull, jaw) and in the long bones. In spine it involves frequently first vertebral bodies to extend in the posterior arch. In the same human species are described multiple primary hemangiomas of bone where the single lesions have all the appearance and clinical characteristics of individual hemangiomas. Less is reported about lymphangiomas in animals and not covered here.

Symptoms When Hemangiomas are symptomatic they manifest themselves clinically with a mild painful symptomatology. Larger lesions that provoke bone resorption can cause pathological fracture. Sometimes the neoplastic disease may be accompanied by the appearance of peripheral osteolytic areas due to pronounced osteoclastic activation stimulated by tumor endothelial cells. Hemangioma has a low rate of local recurrence and there are no data related to progression to malignancy for hemangiosarcoma.

Gross Pathology General features of hemangioma resemble the conventional gross appearance of hemangiomas of skin and in extraskeletal soft tissues. Angiomatous tissues are soft, dark red sponge-like, and easily bleed. Hemangioma of bone usually is localized in medullary cavity and may cause expansion of the cortical surface. During tumor growth the cortex may undergo tumor-induced resorption that can predispose the bone to a pathological fracture.

Cytology Features and Histopathology Hemangioma of the soft tissues and also of bone is composed of an abnormal network of vessels that microscopically is classified as a capillary hemangioma with a pattern composed of small capillaries or a cavernous hemangioma composed of larger vessels. Capillary and cavernous hemangioma represent morphologic varieties of the same tumor. Aspirates from these tumors are infrequently attempted and smears are usually characterized by few groups of spindle cells mixed with different amounts of blood. Histopathology permits the differentiation into capillary and cavernous patterns. Capillary hemangioma consists of small capillary vessels with few erythrocytes in the small lumina and with the larger ones filled with blood. The wall of the capillary vessels is lined with a single flat layer of endothelial cells. Minimal inflammation may be present accompanying vascular-mediated bone resorption and repair of trabecular surfaces of cancellous bone between infiltrating hemangiomatous vessels. In some bone specimens, the vascular tumor may penetrate through areas of cortical bone

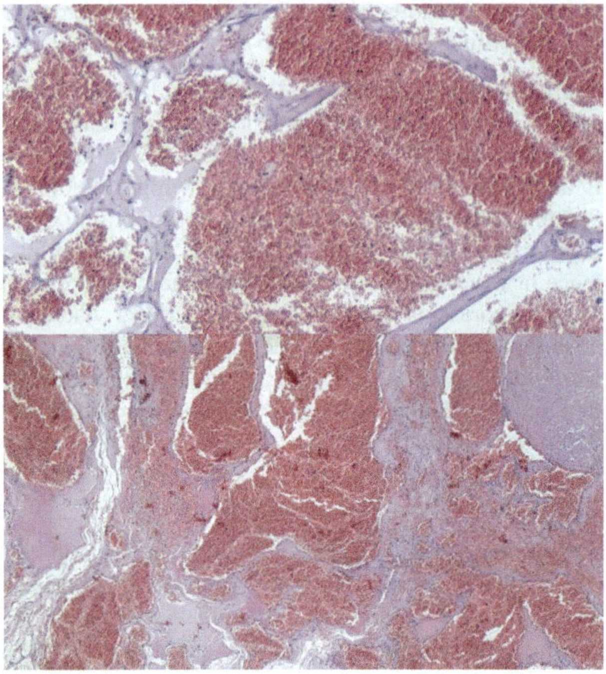

Fig. 2.11 Dog, Border Collie, 13 years old male, typical pattern of cavernous hemangioma (H&E, 4x and 10x, from author collection)

resorption of the host bone of paired long bones and extend into the intervening soft tissue to involve the surface of an adjacent bone. The hemangiomatous cells usually have a flattened cytoplasm with small oval nuclei showing chromatin from fine granular to condensed and small nucleolus not always visible. In more active proliferating cases capillaries may be more packed and the cells may show plump to roundish nuclei. Atypical mitoses are very rare. Usually, the wall of capillary hemangioma of the bone is composed also of a delicate lamina of connective fibers as collagen and reticulin and the affected cancellous bone may vary from being resorbed to being sclerotic. Cavernous hemangioma is differentiated from a capillary hemangioma but having an irregular pattern of large ovoid cavities also filled with blood and by flat mature endothelium and collagen. The histological diagnosis is often easy. The biopsy procedure needs special attention because it may cause intense bleeding. Angiographic investigations prior to biopsy and to use fine needles for sampling should be considered before using trocars or incisional biopsy techniques. Cases of hemangioma of the spine are usually asymptomatic when they develop in young animals. These tumors rarely become aggressive. Malignant transformation does not occur and only local bone tissue may be involved and damaged by the tumor growth and expansion. Prognosis is frequently excellent. In case of diagnostic doubts immunohistochemistry represents an important support for providing a clear and definitive diagnosis. Cells lining vascular cavities stain uniformly for endothelial cells markers such as Factor VIII antigen, CD34, CD31, FLI-1 (Fig. 2.11).

Hemangioma

Definition

Benign vascular tumor of bone composed by well-differentiated vessels with endothelial differentiation.

Image Findings

Classical radiographical findings of vertebral hemangioma are called "striated vertebra" because composed of parallel coarse and vertical bone trabeculation. Compared with other vertebral bodies is possible to identify honeycomb spongiosa with marked trabeculae and frequently intact cortex. It may be present a slight enlargement of the involved bone especially when in the vertebral site. When hemangioma is localized in the skull or long bones it may show a typical "starbust" pattern with spicules of bone in web-like trabecular pattern with also coarse trabeculations surrounded by tumor and that give other typical aspect similar to "corduroy cloth."

To demonstrate the presence of fatty components in the tumor, probably MR investigation is the best way.

Image findings of vertebral hemangiomas could also reveal close to thickened trabeculae areas of round lesions with fats that give an image finding of polka-dot pattern.

Macroscopic features

Angiomatous tissues are soft, dark red sponge-like, and bleeding.

Microscopic pathology

Capillary or cavernous, with lumina lined of single layer of uniform, flat endothelial cells and filled with different amount of blood.

2.11.8 Table Comparing Features of Selected Benign and Malignant Tumors (Table 2.2)

Table 2.2 Main frequent localization and general features between a selection of benign and malignant bone lesions and tumors

	Most frequent localization	Cellularity	Cytological features
Fibrous dysplasia	Long bones, ribs, craniofacial bones	Low	Cytoplasm: Spindle cells Nuclei: Slightly elongated oval-shaped nuclei
Osteoblastoma	Long bones, vertebrae, hip bone	Reasonable moderate	Cytoplasm: Osteoblasts, spindle cells, bone matrix. Nuclei: Eccentric location
Ossifying fibroma	Long bones	Low	Cytoplasm: Few spindle cells Nuclei: Few spindle shaped
Low-grade osteosarcoma	Metaphyseal central or parosteal long bones	Moderate	Cytoplasm: Osteoblasts with bone matrix and osteoid fragments Nuclei: Oval-shaped
High-grade osteosarcoma	Metaphyseal or diaphyseal localization in long bones	High	Cytoplasm: Pleomorphic cells with high degree of anisocytosis. Presence of different types of cells, including also osteoclast-like giant cells. Nuclei: High level of atypia and anisokaryosis, marked nucleoli, multinucleated cells, atypical mitoses.

2.12 Malignant Tumors of Bone

2.12.1 Historical Discussion: Spectrum of Premalignant Lesions, Invasive but Non-Metastatic Tumors, Low-Grade to Highly Malignant Bone Tumors

Pathologist has a great deal of difficulties to diagnose bone tumors, benign or malignant. In the general considerations that contemplate bone tumors of all species I would like to mention what E. Uehlinger reported in 1976 in his text entitled Primary Malignancy, Secondary Malignancy, and Semi-malignancy of bone tumors: at the least judgement, human beings were divided into benign and malignant Christians. However, it soon became evident that the two-class system was insufficient and it was necessary to introduce a third category. This third group was reserved for Christians who would have a chance of becoming clean in purgatory and entering Paradise later. Experience with the biologic classification of tumors was similar: between the benign and malignant types there is an intermediate group. In English literature these intermediate bone tumors are called sarcomas of low grade of malignancy. The German language prefers the term semi-malignancy. The two terms do not have quite the same meaning, but are overlapping. Zollinger introduces the term of semi-malignancy sarcoma as characterized by local invasive and destructive growth; they have a tendency to recur but do not metastasize.

Bone tumors are rare especially in Veterinary Medicine and biopsy frequently is not well performed without take adequate material and, fundamentally, the tissue is poorly processed and errors in biopsy falling in several categories.

Neoplastic and non-neoplastic bone-forming lesions usually are found both in periosteal soft tissues and in bone organs themselves. An aspiration biopsy in radiographically dense heavily ossified lesions generally results in hypocellular or acellular aspirates and may be of nondiagnostic value. A bone needle or trephine core biopsy in that same radio dense bone lesion is also generally nondiagnostic. Diagnosis of bone tumors is not always easy since some traumatic, infectious, and non-neoplastic bone lesions including benign bone tumors can share overlapping radiographic and histological features with malignant bone tumors. Cytohistological investigations also have their unique difficulties in tumor recognition and require a battery of special types of stains. The collected material is then deposited on a slide on which the smear will then be prepared which will finally be dried, colored, and examined under the microscope.

An alternative to this technique is represented by impression exams, where the slide is placed on top of the neoplasm, or on a section of it, and the cells taken are then left to air dry on the slide and then stained. These exams require a high level of experience and deep knowledge of the anatomo-histopathological nature of tumors.

In general, the cytologies of tumors can contain different cellular findings which, in addition to neoplastic cells, can be represented by mixed inflammatory elements (neutrophil granulocytes, eosinophilic granulocytes, lymphocytes, macrophages, plasma cells, etc.) and hyperplastic.

Tumors are cytologically made up of generally monomorphic components, with clear signs of anisocytosis (larger and pleomorphic neoplastic cells), with variable nucleus/cytoplasm ratio, but certainly higher than normal with the exception of some sarcomas and other lymphoid neoplasms.

Cytologically, the degree of malignancy of a neoplasm also takes into strong consideration the alterations that can be highlighted at the nuclear level where the malignant cells contain fine chromatin or massed in a disordered way, the nucleoli appear large and in greater numbers than one per cell and where they can also be present atypical mitotic figures. Cytoplasmic alterations, on the other hand, give important indications on the degree of differentiation of the neoplasm, but unlike the data provided by nuclear modifications, they do not give particular references as regards the degree of malignancy (Table 2.3).

Detailed evaluation of the results of clinical findings, radiologic, CT, and MRI imaging when available, and histopathology often with special stains are useful in diagnosing some cases of a suspected malignant tumor of bone. For many years veterinarians have been resisting the use of modern modalities indicated above for the use of new imaging and histopathology techniques to provide a definitive diagnosis for a bone lesion. These evaluation procedures indicated above are

Table 2.3 Cytological differences between malignant tumors of epithelial origin and malignant tumors of mesenchymal origin

Carcinomas	Sarcomas
Good degree of exfoliation: Easy collection of cells	Low degree of exfoliation: More difficult to take significant samples, especially if from hard tissues such as bone.
Tend to form cellular structures organized in clusters of arranged cells	Generally isolated cells
Cells always have a rounded, oval, or polygonal shape	Cells with variable shape, frequently fusiform and polyhedral
Cell margins are always well defined	Cell margins are generally poorly defined

currently available for obtaining a definitive diagnosis and should be considered because reports have repeatedly confirmed that without the use of those modalities incorrect diagnoses have unfortunately occurred. Some forms of osteomyelitis, for example, can have remarkably similar radiographic images as those images formed by some malignant tumors. Such serious diagnostic mistakes compromise the life of the animals being examined. Biopsy without prior radiographic evaluation of the bone lesion often may lead to mistakes since the bone site offering the best sample of the disturbance may not have been sampled resulting in a nondiagnostic biopsy sample. A "blind" biopsy, i.e. one without guidance from a radiographic observation may not recognize such other bone lesions as chondroma and chondrosarcoma, telangiectatic osteosarcoma and aneurysmal bone cyst. Therefore, all Veterinary pathologists and veterinary clinicians when examining an animal with suspect of bone tumor, they should recall that many non-tumor lesions may radiographically mimic a tumor or pseudotumor.

Pseudotumor represents a lesion that has morphologically different variations of normal tissues and consists of their embryonal precursors. Mesoderm of the developing fetus gives rise to all connective tissues in an animal. In blood vascular and in lymphatic vascular lesions, all their endothelial lining cells originate from primitive undifferentiated mesenchymal elements with stellate-shaped cytoplasm and round central nucleus.

Bones of the skeleton arise from different subtypes of mesenchymal tissue that give rise to cellular precursors that form bone, cartilage, fibrous and adipose tissue, blood, vessels, and bone marrow stroma. Bone marrow and hematopoietic tissue are formed in the yolk sac. The hematopoietic cells arrive via circulation to be housed in the bone marrow of the medullary cavity during formation of the skeleton. Therefore, the supportive structural elements of a bone organ skeleton are formed by and maintained by osteoblasts, osteocytes, and osteoclasts that are derived from bone marrow, chondroblasts and chondrocytes, fibroblasts, and fibrocytes.

Medullary cavity of the bone organ contains mesenchymal derived tissues whose primitive spindle stromal cells form reticulin and collagen fibers that host hematopoietic bone marrow elements (of yolk sac origin, e.g., precursors of erythroid, myeloid, and lymphoid lines), lipoblasts and lipocytes, and blood vascular structures. Note: mesenchymal stem cells from each of the subtypes of mesenchyme are left in situ and can be the cell of origin for tumors of those types of tissue.

Extraosseous tissue of the skeleton also arises from mesodermal tissues derived from fetal mesenchyme. These include precursors of skeletal muscle, fascia within muscle organs, intermuscular fascia, vascular and lymphatic precursors, tendons and ligaments and their supporting connecting tissue, joint capsules and synovium as well as fibrous tissue of the dermis and subcutaneous adipose precursor tissue elements. Note: as in the case of bone organs, mesenchymal stem cells from each of the subtypes of mesenchyme for these extraosseous tissues are left in situ and can be the cell of origin for tumors arising in these seemingly disparate types of tissues.

Therefore, stem cells left behind in any one of these subtypes of mesenchymal "rests" of mesoderm origin can undergo neoplastic transformation and form an example of a benign tumor or give rise to an example of one of a spectrum of low- to high-grade malignant tumor types of mesenchyme-derived bone or extraskeletal tumors.

A multimodal approach to reach the diagnosis of malignant tumor of bone utilizing a gradual systematic evaluation of clinical, radiographic, macroscopic, and microscopic findings and their correlation, and when necessary biomolecular testing.

Most animals with general malignant tumors exhibit normal metabolism of minerals of bone and hypercalcemia is the most frequent described data in a wide variety of primary or metastatic cancers with or without bone metastases. Hypercalcemia according to a study by Woodard in 1953, occurred in 10% of cases with radiological evidence of bone destruction by metastases. Some of these tumors can produce a parathyroid-like substance that can induce a syndrome of hypercalcemia and hypophosphatemia. Other tumors as like as breast cancer may produce sterols with effects of osteolytic action resulting in a chemical syndrome with hypercalcemia not associated with hypophosphatemia (with normal of moderate elevated concentration of plasma phosphate).

The general system of classification is histogenetic. Tumors are classified on the basis of the predominant cell type, and cellular origin and matrix formation. Following this nomenclature an osteoma and an osteosarcoma are tumors formed by neoplastic osteoblasts-osteocytes. A fibroma and fibrosarcoma are tumors formed by neoplastic fibroblasts and fibrocytes. These and the other primary bone tumors are diagnosed according to their histogenesis and histological grade. WHO classification further divided primary sarcomas of bone into two big groups: central bone tumors and peripheral bone tumors.

Malignant tumors of bone involve many animal species. The dog has the highest incidence of malignant bone tumors in domestic animals followed by the domestic cat. Primary sarcomas of bone mainly affect large breeds dogs. Long bones of the skeleton are the sites of greatest frequency for bone tumors. Among all bone tumors, osteosarcoma is the most frequent. Heterogeneity in histologic subcategories of the malignant bone tumor occurs both in animal and human species. One explanation is that following malignant transformation of the osteogenic stem cells left during fetal development, the initial malignant osteoblast as it begins to divide may only replicate of itself, e.g. produce only clones of osteoblastic osteosarcoma. However, neoplastic osteoblastic stem cells can give rise to a clone of mature neoplastic osteoblasts that can form just one of the other different osteosarcoma histotypes, e.g. chondroblastic osteosarcoma, fibroblastic osteosarcoma or giant cell variant of osteosarcoma. Alternatively, the malignant osteogenic stem cells may form clones that are an admixture of these subcategories of osteosarcoma. This concept explains why in multiple biopsies from a single tumor and admixture of these subtypes of osteosarcoma may be recognized. All subtypes apparently have the same biological behavior with a high metastatic potential.

To stage tumors is really important to classify tumors in several groups with similar behavior and prognosis although sometimes very approximately because a complete and very accurate staging system still does not exist.

2.12.2 Barriers to Tumor Expansion; Tumor Locations in a Bone

The general evaluation criteria of sarcomas in general and applicable to veterinary medicine result in years of documented results that are now part of the specialist bibliographic heritage. However, there are still clinicopathological factors inherent in various biological aspects of these tumors that should still be investigated and for this reason we consider it appropriate to report the extensive study on mesenchymal tumors published by Enneking in 1980 and 1983 and relating to the staging system of bone tumors in human medicine based on how the host grows and reacts, also considering the anatomical compartments of the human organism. In particular I would like to report what Enneking established for anatomical compartments in correlation with the growth of malignant tumors such as osteosarcoma. Anatomical compartments represent a structure or a space bordered by natural hedges delimiting the neoplastic growth and extension. These are represented by cortical bone, articular cartilage, joint capsule, fascia, tendons, and tendons sheaths. Extracompartmental other tissues are represented by fat and interstitial areolar tissue located outside the compartments and neuro-vascular structures. Cortical bone or fascia can be involved and breach through vascular structures. Other anatomic barriers are represented by diaphyseal cortex that can restrain the growth by pushing or permeating of a bone tumor. Also, periosteum a thinner barrier than the diaphyseal cortex can increase resistance to the tumor extension by an increase of periosteal production of osteoblasts and bone. The most resistant anatomical barrier, on the other hand, is represented by the articular cartilage which by its nature offers a strong resistance to neoplastic expansion also supported by the fact that having no vascular structures it is also more difficult to perforate by malignant cells (Figs. 2.12 and 2.13).

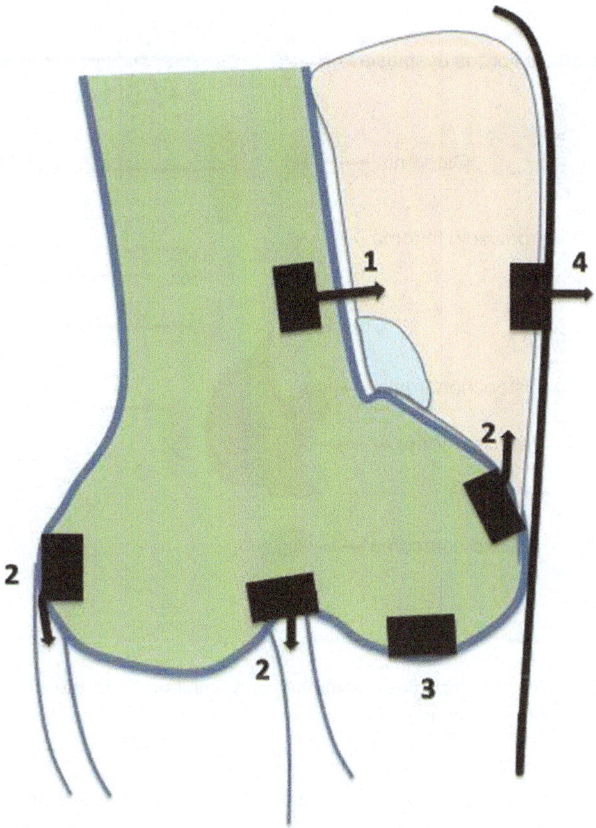

Fig. 2.12 Anatomical natural barriers to expansion of the tumor – 1) Diaphyseal cortex and periosteum. 2) Meta-epiphyseal cortex thin cortex. 3) Articular cartilage. 4) Aponeuroses and fascia (from: Mario Campanacci, modified by the author—Bone and soft tissue tumors)

2.12.3 Osteosarcoma

Definition and Pathogenesis Osteosarcoma is a high-grade malignant tumor of bone composed of mesenchymal cells (malignant osteoblasts) able to produce immature woven bone and osteoid matrix. Primary osteosarcoma of a bone organ arises in the mesenchymal tissue that forms the cortical bone or is present in the medullary cavity of any bone organ. Osteosarcoma seems to affect rapid growing bones more frequently.

For osteosarcoma several potential etiopathogenetic factors have been identified as primary factors or co-factors. These are represented by chemical, physical, and biological agents and described mainly as environmental factors.

According to the World Health Organization, OS of bone is classified into eight subtypes with distinct biologic behaviors and clinical outcomes: conventional, telangiectatic, small cell, low-grade central, secondary, parosteal, periosteal, and high-grade surface. However, in veterinary medicine, examples of these categories which also occur in animals are currently not being recognized and osteosarcomas

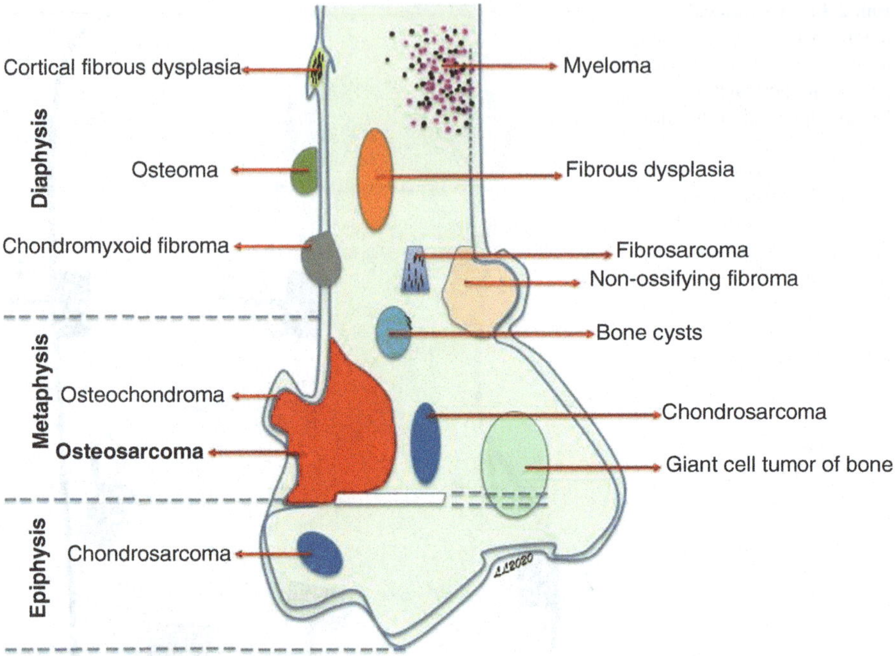

Fig. 2.13 Epidemiological graphic representation of the most frequent localization of bone tumors in canine long bone

animals are classified only into three categories: central, the most common category of osteosarcoma, and two rare parosteal and periosteal categories of osteosarcoma. Examples of the latter two rare categories of osteosarcoma are sometimes described in the veterinary literature as forms of primary extraskeletal osteosarcoma. The diagnostic feature of osteosarcoma is the production of osteoid and tumor bone, i.e. calcified osteoid matrix that has embedded osteogenic cells having cytologic features of malignancy. Probably plain radiographs is the first best diagnostic step for diagnostic suspect of osteosarcoma because it can describe significant features as sun burst appearance, or Codman's triangle, destructive growth, etc. Radiographs can also detect metastasis, especially to the lungs. No specific diagnostic stains are useful in particular but biochemical markers such as alkaline phosphatase or lactate dehydrogenase can give important data for prognosis and response to treatments. Central osteosarcoma demonstrates a high degree of malignancy and metastasizes early in the developmental course of the bone lesion and in some cases has metastasized to the lungs prior to recognition of clinical signs. Some of the comparative age-related biological characteristics of osteosarcoma between humans and animals are interesting. In humans two peaks of occurrence are described. The largest peak occurs in young patients between 10 and 20 years old. The second group consists of osteosarcoma arising in Northern Europeans secondarily to predisposing diseases such as Paget's disease of bone or Li–Fraumeni syndrome.

Paget's osteitis deformans ("Paget bone") is a human bone dysplasia of unknown causes affecting older individuals. Li–Fraumeni syndrome is caused by an autosomal dominant inherited disorder that causes an alteration of expression of the p53 protein. This tumor suppressor protein and recessive oncogene increase the risk (greater than 25 times) of developing one or more tumors in humans. No comparable predisposing tissue factors have currently been recognized as tumor promoting factors in dogs or other domestic animals.

Our previous study is in agreement with other published epidemiological data regarding bone tumors in the canine species. These studies recognize a prevalence of bone tumor frequency in large and giant breeds. These breeds include Great Dane, Saint Bernard, Rottweiler, German Shepherd, Boxer, Irish Setter, Doberman, Golden Retriever. These dogs have a mean age of 7–8 years but also include a few young animals. The age range for diagnosis of osteosarcoma in these epidemiological studies is from 3 months to very old dogs. Some data report that body weight is related to the frequency of occurrence of osteosarcoma in dogs. Canine osteosarcoma which has a 1% frequency in dogs with body weight lower than 10 Kg. An 8% of frequency of osteosarcomas was reported in dogs with body weight ranging from 20 and 35 Kg. A 60% frequency was recorded in dogs weighing more of 35% kg. In 1998, Ru et al. reported that increased height was a more significant factor than increased weight in the occurrence of osteosarcoma. In our diagnostic experience and in our previous studies of canine osteosarcoma, we have found that osteosarcoma is more common in male than female dogs.

Literature reports data for the other animal species. Osteosarcomas represent the most common bone tumor type in cats comprising around 75% of all feline primary tumors of bone. In a study performed in 2000 by Heldmann E. et al., they reported 145 cases of feline osteosarcomas from 1990 to 1995 reported. The mean age of occurrence was 8.03 yrs. ($\pm$ 4.01) for cats with appendicular osteosarcoma and of 10.41 years ($\pm$3.06) for cats with axial osteosarcoma. Males were more often affected than females. Domestic shorthair had the highest frequency of all feline breeds for development of osteosarcoma. About 10% of all cases of osteosarcoma diagnosed in cats were diagnosed as extraskeletal osteosarcoma.

2.12.3.1 Central Osteosarcoma (Conventional Osteosarcoma, Classic Osteosarcoma)

Frequently also called as osteogenic sarcoma is a high-grade malignant neoplasm with malignant osteoblasts producing osteoid. Conventional osteosarcoma is usually abbreviated with the term OSA and it is representing the most common primary tumor of bone of animals that usually arise de novo also without a well-known predisposing condition. This tumor is characterized by all potential variability in osteoblasts differentiation in terms of osteolysis and osteogenesis. For the high grade of malignancy of osteosarcoma, it frequently metastasizes early. The term secondary osteosarcoma describes all forms of this tumor that arise in diseased or predisposed bone tissue. The most frequent cases of diseased or predisposed bone are represented

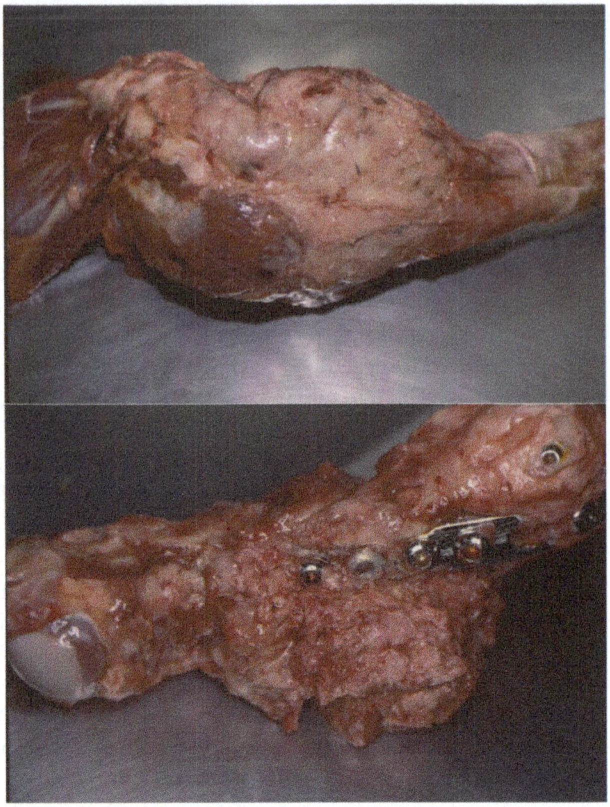

Fig. 2.14 Canine osteosarcoma arising at a metal plate, surgically positioned to repair a previous fracture. Dog (from author collection)

in veterinary medicine by trauma, orthopedic implants (foreign bodies), specific genetic abnormalities, chemotherapy, and radiotherapy. The genetic susceptibility in human OSA is associated with heritable cancer syndromes (Fig. 2.14).

Epidemiology and Localization OSA usually present with palpable mass, swelling, and pain and often is associated with pathologic fracture also because the classic one is an intramedullary destructive mass that extending in the cortex producing a type of periosteal reaction with production of soft tissue mass called Codman triangle described first by Ribbert in 1914 and named later by Ernest Amory Codman after description of similar lesions for Ewing sarcoma (Fig. 2.15).

The production of osteoid or immature bone represents the main feature of osteosarcomas and of the malignant osteoblasts that constitute them. The production-distribution of osteoid is not always constant and regular and can frequently also be associated with the presence of other cellular elements such as fibroblasts, chondroblasts, and multinucleated osteoclast-like giant cells. Most osteosarcomas arising in the internal medullary cavity and subsequently invading the cortex and surrounding areas. Osteosarcoma is a highly invasive tumor also in the surrounding normal tissues and this high degree of infiltrative distribution represents

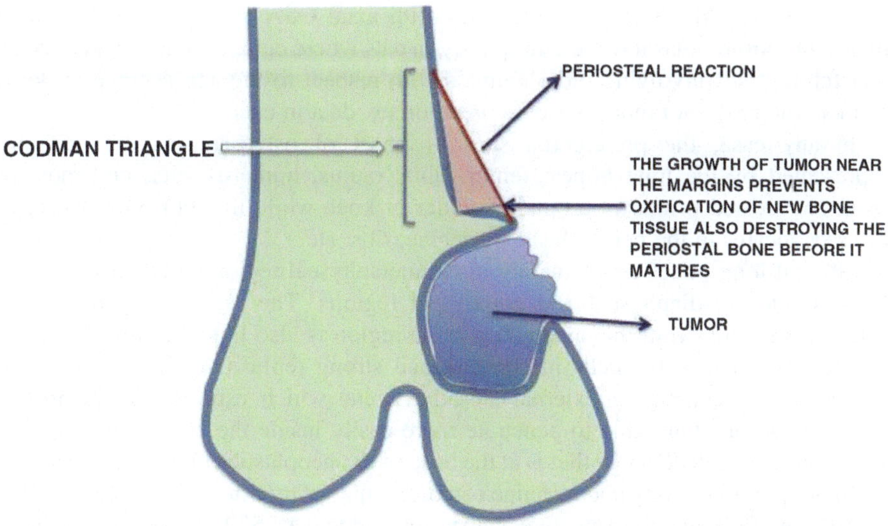

Fig. 2.15 Graphic representation of the peculiar bone lesions, characterizing the formation of Codman's triangle

an important diagnostic feature and, together with other clinic-pathological aspects that we will discuss later, also of an important prognostic factor. Osteosarcoma usually presents a dense and radio-opaque bone pattern subsequently to high grade of mineralization of the mass.

In animals osteosarcoma represents the most frequent malignant bone tumor that accounts for 80–85% in dogs and about 70–75% in cats. There are few other data available for other species where OSA is considered quite rare. In animals OSA occurs most frequently in old or middle-aged dogs with descriptions in the literature of primitive cases of OSA including dogs of 3 months of age up to cases in very old animals (median age 7.0–7.5 years). In August 2020, a group of scientists from Canada and Japan has described in The Lancet Oncology the first case of osteosarcoma in a dinosaur after a multimodal diagnosis application.

In human species OSA has two peaks of occurrence where the first and larger involve patients with a range of age from 10 to 20 years, and it represents the 60% of all human osteosarcomas, and a secondary peak that involves mostly late adult.

Regarding to the distribution-frequency of the tumor in animals by breed, it has long been demonstrated that in the canine species the most affected breeds are represented by large and giant breeds: German Shepherd, Great Dane, Saint Bernard, Boxer, Rottweiler, Golden Retriever, etc.

In a published paper from 1997, Delisle and Devauchelle indicate that dog OSA has a frequency of 1% in subjects weighing less than 10 kg, 8% in those weighing between 20 and 35 kg, and 60% in dogs weighing more than 35 kg compared to all cases of primary OS diagnosed in the canine species. There is a different distribution of OSA for sex and breeds where males seem to be involved with light difference versus females where otherwise females with axial OSA are more involved than males. Osteosarcomas have less frequency in cats where epidemiological data report that feline cases involving older animals with respect to the age average in dogs. Epidemiology do not report specific breed or sex data in cats.

In any case, the preferential sites of onset of osteosarcoma are certainly represented by the long bones (femur, tibia, radius, humerus, etc.) and most of them are located primarily around shoulder or knee while the remaining minority of cases locates in the pelvis, head, vertebrae, ribs, etc.

OSA of long bones arises mainly in the metaphyseal region and then spreads to the neighboring diaphyseal and epiphyseal regions. The reason why the tumor originates mainly from this anatomical bone region is also linked to the high local cell turnover where the cells in constant and strong replicative activity are more susceptible to the action of external causative agents which, especially at the time of cell division, they are able to penetrate more easily inside the nucleus causing the irreversible genetic damage that is at the base of the neoplastic cellular modification.

In our previous study about canine osteosarcoma we reported, from the year 1991 to the year 2000, in a retrospective analysis considered 12,520 biopsies and surgical samples from dogs of different breeds, sex, and ages, 68 primary malignant bone tumors, of which 59 (86.76%) represented by osteosarcomas classified as proposed by the WHO in 1994. The most frequent subtype was osteoblastic with 39 cases (66.1%), followed by from the fibroblastic (9 cases equal to 15.25%), from the giant cell one (4 cases equal to 6.78%), from the poorly differentiated model with 3 cases (5.09%), and finally from the chondroblastic and telangiectatic types with 2 cases (3.39%).

Symptoms and Clinical/Serological Findings In case of osteosarcoma the first symptom is usually represented by pain, frequently referred to trauma. In few weeks pain increase with the appearance of a painful swelling. Advanced clinical signs can be also the high body temperature and limited joint motion. Several cases of osteolytic osteosarcoma pathologic fracture may occur.

Serological investigations of patients with osteosarcoma may indicate elevated values of alkaline phosphatase (ALP) and of lactate dehydrogenase (LDH).

Serum alkaline phosphatase concentration represents a prognostic factor for osteosarcoma.

In dogs the main causes that determine an increase in the serum ALP activity are biliary stasis, treatments with corticosteroids, tumors, acute pancreatitis, some forms of liver diseases, while in the cat, where the serum ALP levels are certainly lower, the variations more detectable are frequently caused by kidney problems or by pathological bone remodeling factors. As a prognostic factor it was reported that of bone specific ALP is associated with short survival in dogs with osteosarcoma.

However, is essential to remember that diagnosis of OSA requires a strong collaboration between clinicians, radiologists, and pathologists involved in a triple diagnosis evaluation which include different aspects of the tumor (Fig. 2.16).

Gross Pathology In general, the osteosarcomatous mass is made up of an amorphous mass of tissue that manifests itself as a dense bone swelling that inordinately

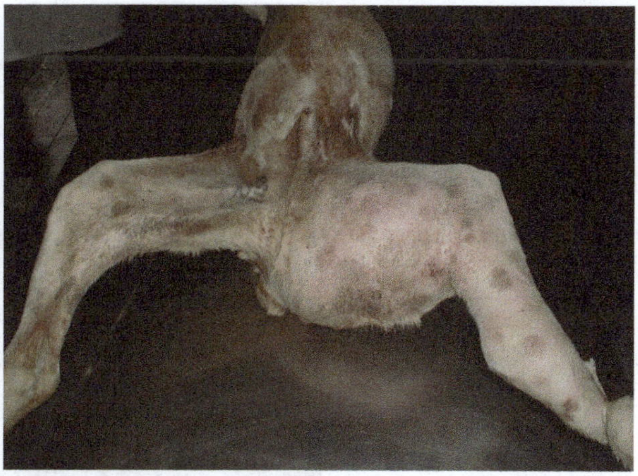

Fig. 2.16 Gross pattern of primitive osteosarcoma in a dog, with swelling of the affected left limb (from author collection)

replaces the normal architecture of the bone. OSA is characterized by permeative and destructive lesion by progressive enlarging mass usually around the metaphysis of long bone (rarely the tumor is localized and confined to the epiphysis) that in 80–90% extend into soft tissue. Gross appearance of OSA is very variable and the tumor tissue appearance depends on various factors such as tumoral matrix production, features of tumor cells, and reactive osteogenesis. Many forms of OSA are lytic, many other productive. In its different morphological and macroscopic aspects, OSA have characteristics of fleshy mass in areas where the matrix is not present, or of more consolidated and rubbery mass in areas where there is production of collagen mixed with fibroblasts. In other cases, OSA can present as a granular or markedly hard tumor when there are highly productive osteogenic areas, or as a cartilaginous or cartilaginous-myxoid tumor when a marked chondroblasts cell component is present. Non-mineralized cartilaginous parts are shiny and gray. Usually, periphery of the mass is less mineralized than central areas. Frequently present on macroscopic examination are represented by the presence of necrosis, hemorrhages and, more rarely, cysts. Is also possible to detect "moth-eaten" osteolytic foci of spongiosa and cortex and pathological fracture through the bone. Usually, the bone cortex is involved by the growth of the tumor which permeated and breach it associated with endosteal and periosteal production of reactive bone with formation of eccentric soft tissue component that dislocate peripherally the periosteum (destruction of cortex). The periosteum is greatly thickened and partially ossified and one important diagnostic feature is also a permeation of bone marrow spacing. Osteosarcoma may growth into joint space, through synovium, along cortical surface or also across joint capsule and tend ligamentous insertion sites.

Rarely tumor cells can be detected in the draining veins around the tumor. Skip metastases involved usually first the same affected bone and appear as early intramedullary metastases in the marrow cavity as ovoid, firm, white-brown nodules near or far from the primitive mass (Fig. 2.17).

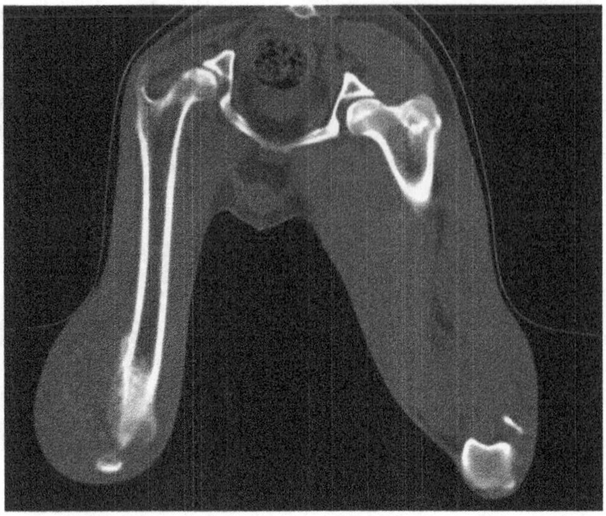

Fig. 2.17 Dog, mixed breed, 2 years old, osteosarcoma. The Axial CT scan reveals an intramedullary osteo-proliferative lesion at the level of the distal epiphyseal and metaphyseal part of the femur. From the lateral periosteal surface develop an expansile mass characterized by a fine granular bone proliferation (Courtesy of Dr. Giovanni Angeli—Department of Veterinary Medicine, University of Perugia)

Cytology Features and Histopathology Fine-needle aspiration and core biopsy are the most used methods to obtain tissue samples for histological examination. OSA can be a histologic mimicker and lead to misdiagnosis especially when examining small biopsies. The related cytologic characteristics may vary with the type of OSA and related to predominance of the specific histotype. OSA is usually more cellular than aspirates from other soft tissue sarcomas.

Important to remember that in several cases of OSA, both in the purely lithic forms (with frequent areas of necrosis and hemorrhages) and in the more properly productive ones (with the formation of compact and very hard bone), it is frequently very difficult to be able to take a significant and cytologically representative sample for the issuance of a certain diagnosis, so in many cases it may be necessary to perform multiple sampling of the lesion.

Smears are usually highly cellular and can contain a background of red blood cells. The cytological features of osteosarcoma consist of a prevalence of single osteoblastic cells, isolated or in clusters, that show different degrees of atypia which can vary from an oval or plasmacytoid type with eccentric nuclei in well-differentiated forms. Infrequently, binucleated or multinucleated osteoblast-like cells can be detected. Coincidently may also be present paranuclear vacuolization and small and dense cytoplasmic granules both expression of the presence of immature osteoid matrix. In many samples, multinucleated osteoclast-like cells can also be detected frequently associated to spindle cells and fragments of osteoid, often calcified, cartilage and necrosis. Respect chondroid, the osteoid show an appearance of denser and more homogeneous pattern and is always associated at the presence of osteoblasts. Chondroid is more myxoid and more dark with respect to osteoid. Malignant osteoblasts show clear signs of anisokaryosis and it appear enlarged, irregular in shape sometimes folded, cracked and with irregularly furrowed edges and a common cytological relief is represented by the frequent presence of mitotic figures, particularly present in cases of OSA with a high degree of malignancy.

For the cytopathological diagnosis, the use of staining for ALP can also be used to support diagnoses of OS even if it does not allow to differentiate reactive osteoblasts from malignant osteoblasts, but the expression of its marked positivity allows the pathologist to orient himself towards a more certain diagnosis of osteosarcoma.

Although histopathology is strongly influenced by the wide variety of histotypes that can characterize the various forms of OSA the typical feature of it is of a high-grade sarcoma with malignant polyhedral osteoblasts capable of producing osteoid and woven bone. The identification of tumor malignancy is therefore based on the highlighting of cellular and nuclear pleomorphism, on the degree of nuclear and cytoplasmic atypia, on the presence of bizarre mitosis, and markedly hyperchromatic chromatin. The tumoral osseous matrix can show a pattern of small island or form into denser and more intertwined packets messily organized. In most cases of osteosarcoma is predominantly the osteoblastic differentiation in different pattern as minimal, extensive, or focal and it may be mixed with foci of collagen, production of cartilage and/or fibro-histiocytic cellular elements. An admixture of two prevailing elements represents in different proportion the common histologic feature of osteosarcoma. A bone matrix is produced by tumoral cells and it is formed by hyaline and eosinophilic amorphous material surrounded by the malignant cells.

Pleomorphism and atypia involved nuclei and cytoplasm are diagnostic. Occasionally also para-nuclear vacuoles are present associated with small densely dark cytoplasmic granules represent osteoid.

Osteosarcomatous cells have hyperchromatic nuclei, central or eccentric in the cell soma. Mitotic bizarre figures and prominent nucleoli are very common, with different levels of atypia, eosinophilic cytoplasm, and high variability in volume. The osseous matrix amount varies and is irregularly distributed with no ordered trabeculae usually rimmed by osteoblasts.

Low-power magnification of the tumor represents a first fundamental step to identify and evaluate the pattern of the tumor and the degree of involvement of the bone involved: involvement and destruction of the cortex, of the medullary canal, of the periosteum, vascular invasion, and/or invasion of adjacent soft tissues, etc.

In the osteogenic/sclerotic form of osteosarcoma the tumor cells are not as abundant as in the more productive forms and these may be small, with slim nuclei and denser chromatin and with no or only few numbers of bizarre mitoses. Foci of necrosis can be frequently present. The vessels network in the tumor increasing in the osteogenic areas are often represented by dilated sinusoids with discontinuous wall. Multinucleated giant cells are seen frequently (reactive action, especially near areas of hemorrhages and hypervascularity) but mononuclear cells usually predominate.

The epidemiological data present in the literature and relating to osteosarcoma in veterinary medicine report that the dog represents the animal species definitely most affected, followed by the cat and, in significantly less horses. Osteosarcoma affects more frequently middle-aged dogs (7–8 years old), males, and large breed dogs. In dogs and cats generally, most osteosarcoma produce early-stage metastasis after several months if not early detected. Feline osteosarcomas usually show large islands of cartilage.

As also White V.A. and colleagues did in human medicine also to facilitate the diagnosis of osteosarcoma, a group of veterinary pathologists, enclosed Prof. Roy R Pool, in collaboration with the World Health Organization, made in 1994 the histological classification of bone and joint tumor of domestic animals where the classification scheme for osteosarcoma is still currently accepted which subdivided the tumor in:

(a) **Poorly Differentiated**
(b) **Osteoblastic (productive and non-productive)**
(c) **Chondroblastic**
(d) **Fibroblastic**
(e) **Telangiectatic**
(f) **Giant cell type**
(g) **Mixed (not enclosed in WHO classification)**

(a) **Osteosarcoma Poorly Differentiated**: consisting mainly of mesenchymal cells with high grade of anaplasia and pleomorphism and may show fusiform, round, plasmacytoid, spindle, and small with eosinophilic clear cytoplasm. Cell may show different marked signs of anisocytosis and anisokaryosis and different amounts of osteoid and neoplastic woven bone with not organized trabeculae. When unmineralized the bone is eosinophilic and differently when mineralized basophilic. Sometimes it is difficult to distinguish osteoid from other extracellular eosinophilic material, as like as collagen, that appears in different intercellular aggregates more fibrillar and compressed with respect to unmineralized matrix. Most of poorly differentiated osteosarcomas are very highly aggressive and frequently associated to pathological fracture due to severe bone lytic damages (Fig. 2.18).

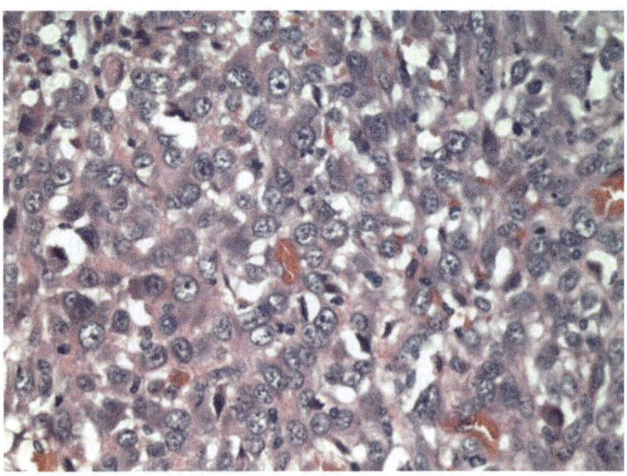

Fig. 2.18 Poorly differentiated osteosarcoma (Courtesy of Dr. Roy R. Pool, Emeritus Prof. UCD and Texas A&M University)

(b) **Osteoblastic Osteosarcoma (Productive and Non-productive)**: Osteosarcoma has a very wide range of histological form and presentation but this form producing osteoid and woven bone is the most frequent form of osteosarcoma in animals and also in humans. Osteoblastic osteosarcoma is formed mostly by tumor osteoblasts and different amount of osteogenic precursor-cells spindle-shaped or plump. The osseous matrix is produced by tumor cells in different amounts and different patterns, from dense sheets to coarse island. The typical chessboard-like distribution of the neoplastic osteoid is one common feature. The different amount of this bone tumor matrix characterized osteoblastic osteosarcoma is divided into two further subcategories: OSA productive with abundant amounts of osteoid and OSA non-productive in case the bone tumor matrix is present in minimal quantities as like as in most of lytic bone neoplastic lesion. Malignant osteoblastic cells are polyhedral cells with hyperchromatic pleomorphic nuclei, with hyperchromatic chromatin. Nuclei are eccentric in a basophilic cytoplasm. Together with bizarre and atypical mitosis these cellular features characterized the first histological criteria of osteoblastic osteosarcoma. At sarcomatous stroma are frequently associated multinucleated giant cells, hemorrhages, calcifications, connective fibrils, and fibrous tissue. The typical aggressiveness of this subtype of osteosarcoma justifies also the lytic bone lesions and the frequent response of the periosteum. The destruction of bone is also the cause of the frequent pathologic fracture. Identification of osteoid structures is not always easy in osteoblastic osteolytic osteosarcoma especially when tumor is highly cellular and can be helpful to investigate it with the Goldner staining technique (Fig. 2.19).

(c) **Chondroblastic Osteosarcoma**: this subtype is characterized by a strong component of malignant chondroblasts and chondrocytes producing also chondroid matrix. The main feature of chondroblastic osteosarcoma is the presence of both chondroid and osteoid matrix. The cartilaginous malignancy of this tumor shows

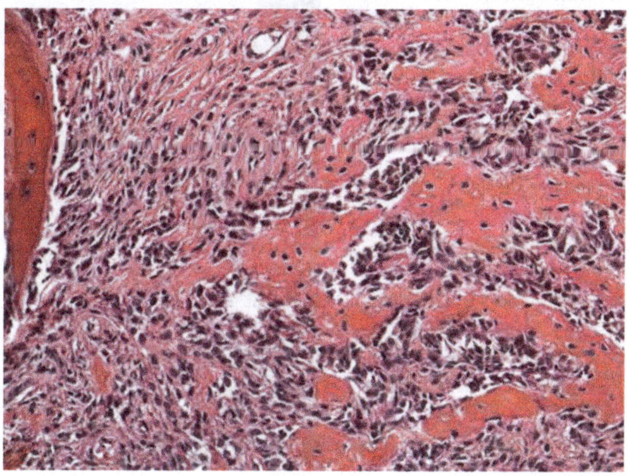

Fig. 2.19 Osteosarcoma (typical) (Courtesy of Dr. Roy R. Pool, Emeritus Prof. UCD and Texas A&M University)

high degree of neoplastic hyaline cartilage also associated with non-cartilaginous derivatives, but with osteoid or bone matrix with various degree of differentiation. The tumor cells are often entrapped in matrix and frequently when chondroid differentiation is present malignant cells are detectable internal at cartilaginous lacunae. The smears from chondroblastic osteosarcoma are hypercellular and, only in rare cases, its shows low cellularity. In association to sarcomatous cells it may be frequent to detect plasmacytoid cells with vacuolated cytoplasm, multinucleated and binucleated cells, and foci of necrosis. This form of tumor represents one of the cases in which a biopsy sampling not performed properly can make difficult the issue of the definitive diagnosis which must always include a differential evaluation in the first instance with chondrosarcoma (Figs. 2.20, 2.21, 2.22, and 2.23).

(d) **Fibroblastic Osteosarcoma:** occasionally osteosarcoma is fibroblastic, similar to fibrosarcoma but with a mixed pattern with neoplastic spindle cells and osteoid-osseous production by the same cells. This pattern is identifying osteosarcoma that generally has more favorable prognosis with respect to the other subtypes. In humans this form represents the 10% of all cases of osteosarcoma and this variant is also defined sclerosing type when particular rich of malignant fibroblasts. Pathologists must take particular attention during the microscopical evaluation of the tumor or its metastases to identify osteoid-osseous matrix that sometimes may be scarce and similar to collagen aggregates which instead are typical of fibrosarcoma. In addition it can be frequent to detect osteoclast-like

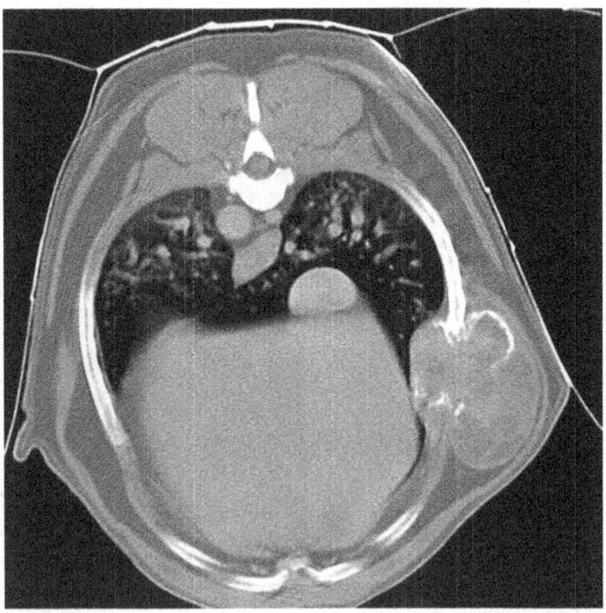

Fig. 2.20 Chondroblastic osteosarcoma, mixed breed, 10 years old, male. The CT axial view of the thoracic wall at the level of the seventh right rib shows an expansile mass developing from the bone part of the rib and is characterized by a wide osteolithic lesion. The aspect of the mass is disomogeneus with an irregular contrast enhancement and a soft tissue-like density. Proliferative perilesional strips of new formed bone are also present (Courtesy of Dr. Giovanni Angeli—Department of Veterinary Medicine, University of Perugia)

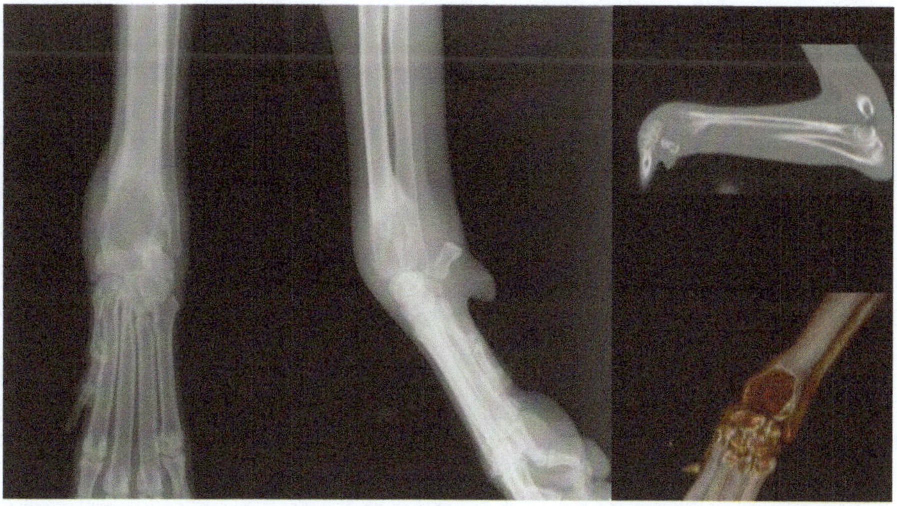

Fig. 2.21 Chondroblastic osteosarcoma in an Abruzzese Shepherd dog, 2 years old, female (S). The cranio-caudal and medio-lateral X-ray projection of the front limb distal part show a well-defined rounded hypointense lesion of the metaphyseal and epiphyseal tract of the radius. A proliferative periosteal production is more evident in the caudal part of the lesion. The computed tomography sagittal reconstruction and the volume rendering demonstrate the high level of bone reabsorption (Courtesy of Dr. Giovanni Angeli—Department of Veterinary Medicine, University of Perugia)

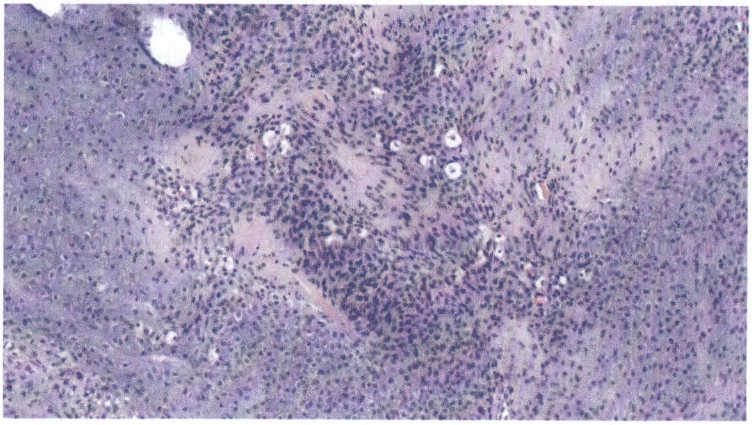

Fig. 2.22 Chondroblastic osteosarcoma, cat, European, 14 years old, F(S) (H&E, from author collection), 10x

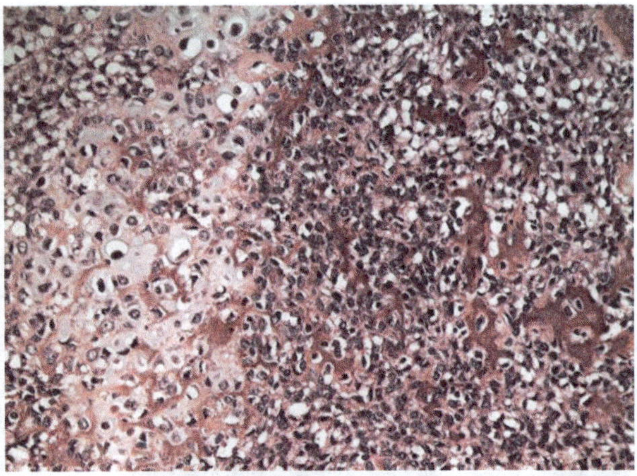

Fig. 2.23 Chondroblastic osteosarcoma (Courtesy of Dr. Roy R. Pool, Emeritus Prof. UCD and Texas A&M University)

reactive giant cells and periosteal reaction with reactive osteogenesis (Figs. 2.24, 2.25, 2.26, and 2.27).

(e) **Telangiectatic Osteosarcoma:** this subtype represents a frequent aggressive tumor of bone characterized by sponge-like structure mixed with diffuse bloody cystic lesions and with immature and scant osteoid. These hemorrhagic and multicystic lesions of blood clots are described clinically in human species as a "bag of blood." Telangiectatic osteosarcoma usually has a fast-aggressive course in all animals' species enclosed the human one. Pathological fracture is frequent due to severe osteolysis and aggressive course. Macroscopic features of telangiectatic osteosarcoma can lead to confuse this tumor with a hemangiosarcoma or an aneurysmal bone cyst and it is represented by a bleeding sponge tumor with multiple blood cavities of variable sizes filled of blood and separated by thin fibrous septa containing malignant neoplastic cells. Necrosis is usually detectable. The neoplastic tissue may represent also the stromal wall and septa and frequently show a very permeative pattern that can reach and destroy also the periosteum and the bone cortex. Cytologically telangiectatic osteosarcoma is often low cellular with a predominant background rich in blood red cells. This can preclude the differential diagnosis with aneurysmal bone cyst that it can be facilitated by the presence of pleomorphic spindle cells and osteoblasts with hyperchromatic nuclei. Histologically telangiectatic osteosarcoma is usually a very highly cellular tumor that under low-power has a similar pattern to aneurysmal bone cyst with multiple large hematic cavities and septa containing also reactive osteoclast-like giant cells. Under high power the malignancy of the cells appear clear and this is almost what is clearly detectable in the prevalent majority of cases of telangiectatic osteosarcoma. Mitotic rate of this tumor is always high and tumorous osteogenesis frequently scarce and focal. Malignant cells that filled the blood-filled cavities and septa result always negative for immunohistochemistry for Factor VIII, a marker of endothelial cells helpful to use for

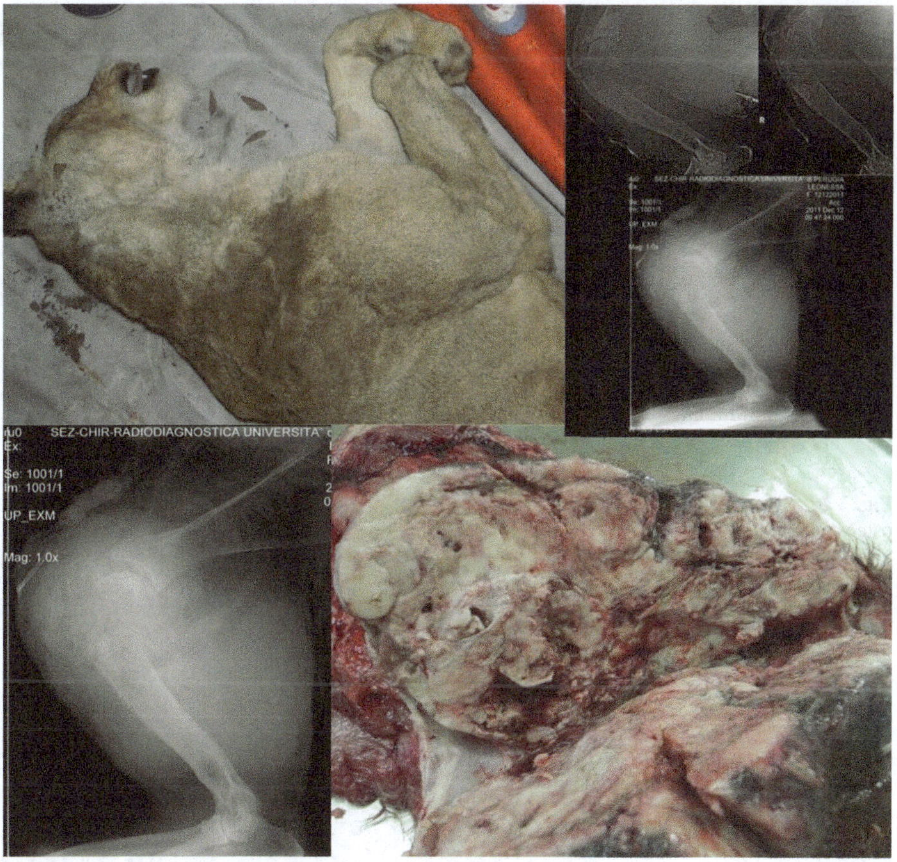

Fig. 2.24 A case of fibroblastic osteosarcoma of the humerus, lion, 16 years old, female. Gross and radiographic pattern (from Leonardi et al. 2014b)

differential diagnosis with hemangioma/hemangiosarcoma. However, in the first instance, the differential diagnosis needs to involve aneurysmal bone cyst. Telangiectatic osteosarcoma contains spindle cells with polymorphic nuclei and abnormal mitoses and irregular and scant neoplastic osteoid, some osteoclastic-like multinucleated giant cells, numerous dilated blood vessels permeating throughout the neoplastic stroma. Necroses can be detected in telangiectatic osteosarcoma but not in aneurysmal bone cysts. Metastases features resembling the primary tumor with many cavities filled with clots and blood (Figs. 2.28 and 2.29).

(f) **Giant Cell Type Osteosarcoma:** this type is characterized by an abundant amount of multinucleated osteoclast-like giant cells originated from the fusion of many malignant osteoblasts. In classic osteosarcoma the presence of multinucleated giant cells is common and the differential diagnosis between

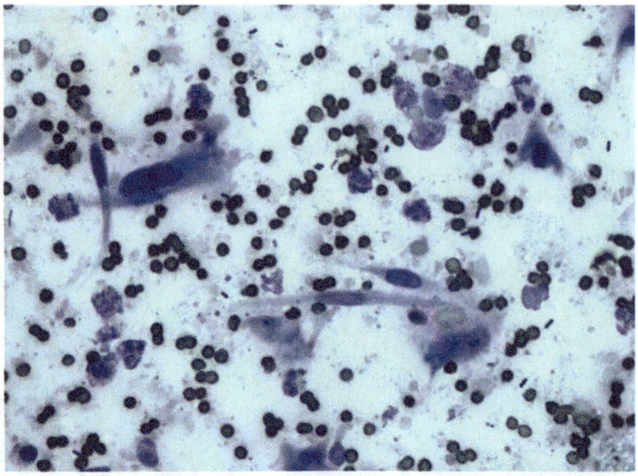

Fig. 2.25 Cytological features of fibroblastic osteosarcoma in a lion, with elongated and fibroblastic-like cells with scant basophilic cytoplasm and elongated nuclei with occasionally prominent nucleoli (from author collection)

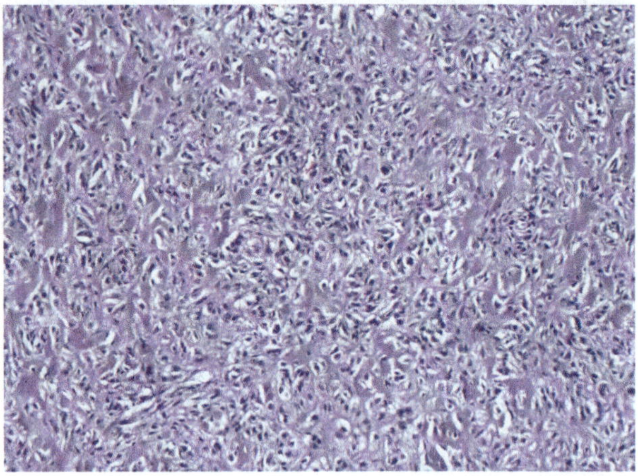

Fig. 2.26 Fibroblastic osteosarcoma in a lion, histological pattern (10x), H&E, from author collection

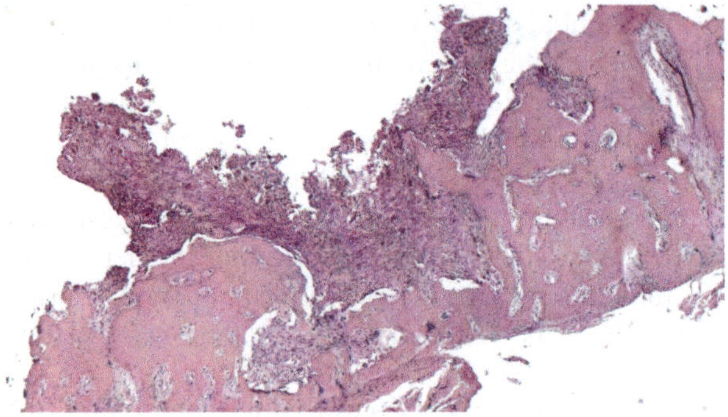

Fig. 2.27 Fibroblastic osteosarcoma, cat, European, 10 years old, male (H&E, from author collection), 4x

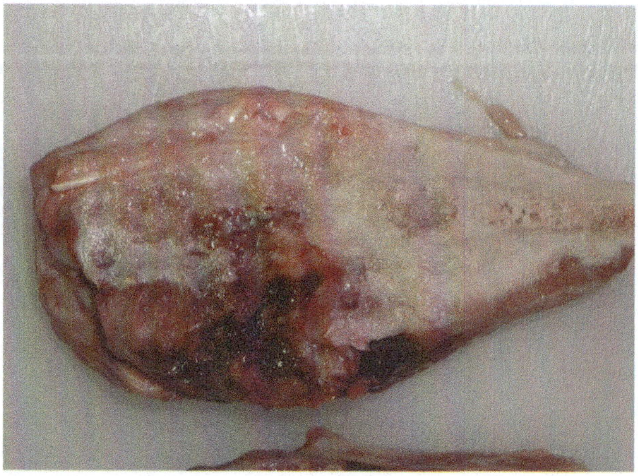

Fig. 2.28 Telangiectatic osteosarcoma in a dog with the aggressive features and invasion of epiphysis and partially of the joint (from author collection)

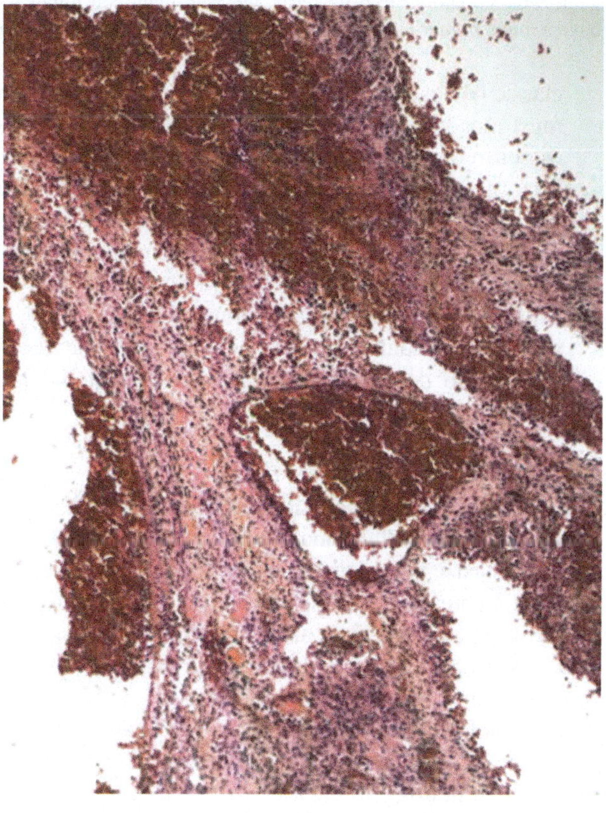

Fig. 2.29 Telangiectatic osteosarcoma (Courtesy of Dr. Roy R. Pool, Emeritus Prof. UCD and Texas A&M University)

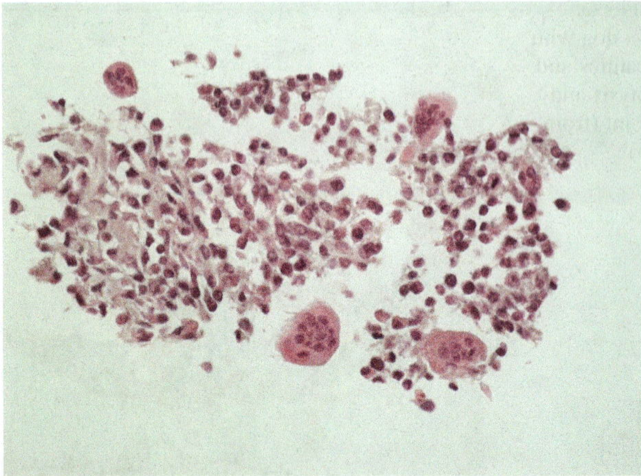

Fig. 2.30 Cytology of giant cell osteosarcoma, dog. Typical pattern containing mononuclear malignant cells with dark and angular nuclei, and multinuclear giant cells and moderate production of mild pink osteoid between mononuclear cells. In giant cell osteosarcomas the multinucleated giant cells are similar to osteoclasts, while in giant cell tumor of bone multinucleated giant cells have a much higher number of nuclei and bigger in volume (from author collection)

classic osteosarcoma, giant cell type osteosarcoma, and giant cell tumor of bone must always be considered. Cytological features of smears from this subtype of osteosarcoma show moderate to high cellularity with a prevalence of malignant osteoblasts, less number of spindle cells, and a significant number of osteoclast-like giant cells that may also contain more of 50 nuclei regular in shape (oval or coffee bean shape) and without clear signs of anaplasia immersed in an abundant cytoplasm. The cytological highlighting of many mononuclear cells (spindle cells and osteoblasts like) with marked signs of atypia orientates the pathologist in the differential diagnosis towards the classic giant cell tumor of bone. Often is possible to detect several fragments of osteoid matrix (Figs. 2.30, 2.31, and 2.32).

Other General Considerations about Central Osteosarcoma

Osteosarcomas in all animals usually have a rapid aggressive local growth course and metastasize very quickly and especially without early diagnoses patients died of metastatic spreads also despite surgical ablation of primary tumor, for the presence of occult micro metastases also at the time of diagnosis. The way of metastasizing is usually hematogenous and occurs primarily to the lungs. Only few cases are described in dogs with skip metastases to the same bone or adjacent one and metastases to regional lymph nodes are equally rare. The process of metastasis in cats appears to have a lower frequency with slightly more favorable prognostic factors than in dogs. Local therapy is characterized by limb salvage wide resection and radiation therapy is used mostly for unresectable tumors. The prognosis of

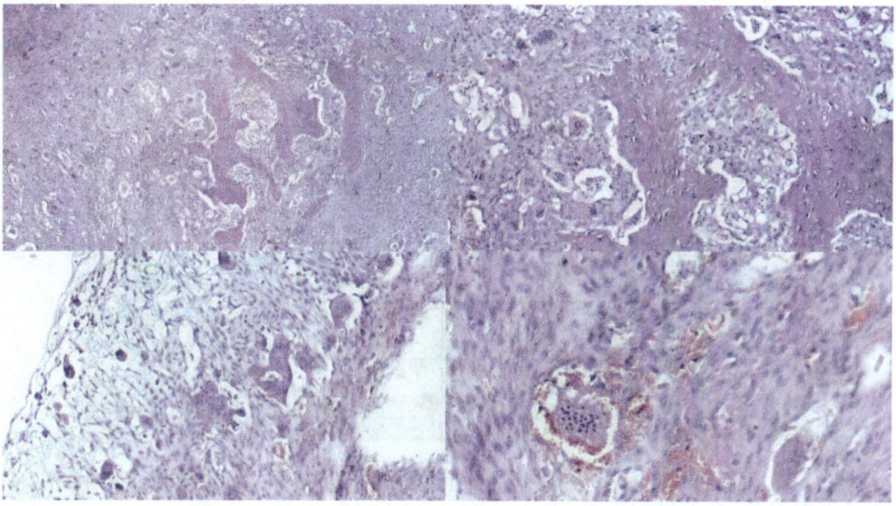

Fig. 2.31 Giant cell osteosarcoma, cat, Maine Coon, 11 years old, female (S). Typical feature of the tumor at several magnifications. (H&E, from author collection)

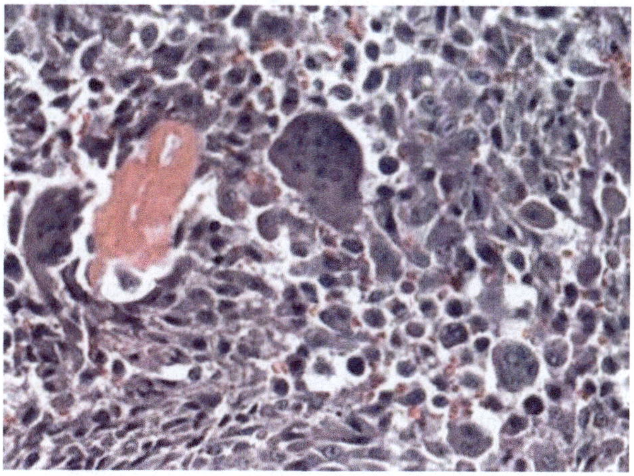

Fig. 2.32 Giant cell type of osteosarcoma (Courtesy of Dr. Roy R. Pool, Emeritus Prof. UCD and Texas A&M University)

animals is influenced by many factors such as age, sex, tumor size and tumor volume, location, and presence of micro and visible metastases, stage and surgical margins.

In humans osteosarcomas pretreated with chemotherapy and with >90% chemotherapy-induced tumor necrosis after complete resection of primary tumor have a prognostic factor of 5-year survival rate of about 80%. When the value of chemotherapy-induced tumor necrosis after complete resection of primary tumor becomes less of 90% then the prognosis becomes poorer.

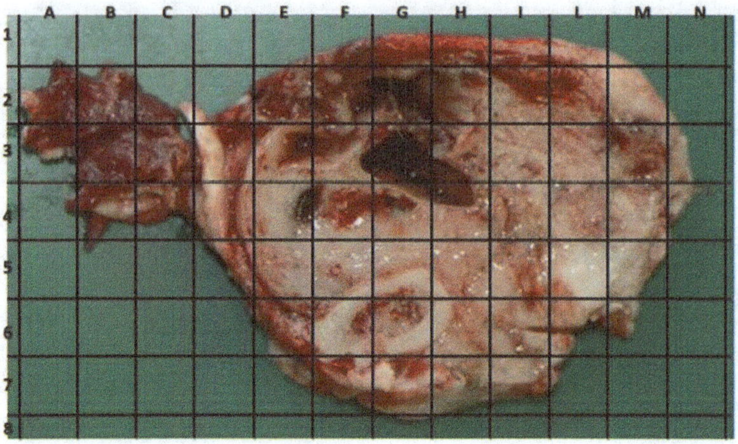

Fig. 2.33 Section full thickness of a case of canine osteosarcoma to evaluate prognostic factors after application of local irradiation therapies

This interesting pre-surgical chemotherapy treatment procedure is histologically evaluated also by the pathologist who, receiving the primary osteosarcomatous mass, totally removed after treatment, and dissected in its major surface. The tissue section is then divided into many small samples which are appropriately numbered and all embedded in paraffin and subjected to histological examination for the quantitative morphological evaluation of the necrosis index produced by the effect of the therapies (Figs. 2.33 and 2.38).

Anyway, the surgical approaches for osteosarcomas are represented by limb salvage or amputation when reconstruction is impossible or pathologic fracture occurs. Adjuvant therapy with preoperative chemotherapy makes a reduction of tumor size, and chemotherapy efficacy can permit to identify the histologic assessment of the tumor and its related necrosis ratio induced by preoperative chemotherapy. The most widely used drugs for the chemotherapy treatment of osteosarcoma are cisplatin, high-dose methotrexate, and ifosfamide.

Unresectable osteosarcoma can be treated with radiation, as like as patients with widely metastatic incurable disease.

Histochemical stains allow also to detect a high content of alkaline phosphatase in osteosarcomatous cells especially in case of patients with elevated serological levels of alkaline phosphatase. The immunophenotype of osteosarcoma is characterized by an immunological profile that lacks of specificity and which can express a series of positivity towards some antigens such as:

Osteocalcin and osteonectin.
RANK and RANKL.
Metalloproteinases (MMP_2 and MMP_9).
S100 protein (especially for cartilaginous areas).
Actin.

Vimentin.
SMA (Smooth Muscle Antibody) and NSE (Neuron-Specific Enolase).
CD99 (Cluster of Differentiation 99, or MIC2 or Single-Chain Type-1 Glycoprotein).

It is important to remember that among the various diagnostic pitfalls of osteosarcomas there is also the one that contemplates the possibility of detecting certain positivity towards keratin and EMA (Epithelial Membrane Antigen).

The Ultrastructure of osteosarcoma cells resembling the features of mesenchymal cells with a dilatation of abundant rough endoplasmic reticulum, eccentric nuclei, and prominent Golgi apparatus. Unlike other neoplastic cells (differential character) osteosarcomas contain collagen fibers in their matrix among which it is possible to detect the presence of focal deposits of hydroxyapatite crystals. These features have not a diagnostic value.

Molecular genetics of osteosarcoma aberrations is well documented in literature with all osteosarcomas that contain clonal chromosomal aberrations especially in dogs and in comparative aspects with humans.

Most common differential diagnoses:

- Osteoblastoma. Aggressive osteoblastoma is a very rare tumor in animals and only few case was reported in a cat and in a pony but there are several features in comparison with osteoblastoma which can guide the pathologist's diagnosis towards an osteosarcoma:
  - Osteosarcoma has a large size more then 5–6 centimeters in diameter.
  - Fast and deep infiltration of normal bony trabeculae. Osteoblastomas do not infiltrate pre-existing lamellar bone structures.
  - Bone deposition by neoplastic cells.
  - Atypical mitotic figures are features of OSA. On the other hand, osteoblastoma shows osteoblasts that may have mitoses but not atypical ones.
  - Moreover, histologically osteoblastoma is composed of structures of tumor bone trabeculae lined by many osteoblasts.
- Chondrosarcoma. Chondroblastic osteosarcoma and mandibular osteosarcoma may have a scarce amount of bone and marked. When tumor with abundant cartilaginous components show high grade of cellular atypia a more well-founded suspicion may be advanced for osteosarcoma. Chondrosarcoma produces chondroid or fibrillar matrix but never osteoid.
- Fibrosarcoma. It is a pure fibroblastic tumor originating from fibroblasts and myofibroblasts and arises from stroma elements and producing spindle cells and no bone or cartilage.
- Giant cell tumor of bone usually has a peripheral reactive woven bone. Respective classic osteosarcoma giant cell tumor of bone has a bone lined by normal osteoblasts instead of atypical tumor cells as like as in osteosarcoma. Giant cell tumor of bone has multinucleated giant cells that contain also up to 100 nuclei.

- Aneurysmal bone cyst similar to telangiectatic osteosarcoma but with cells containing in cyst wall do not have marked signs of atypia as like as in telangiectatic osteosarcoma.

Staging and Grading: There are several factors in human and animal species used to establish a grading for osteosarcomas. Most grading systems of all sarcomas are based on mitotic activity, degree of cellular differentiation (depending on histological type and subtype), and necrosis. These are represented first by the origin of primitive tumor (central, medullary, parosteal, periosteal, etc.), cellular modifications and atypia, mitotic index, necrosis, hemorrhages, neoplastic bone and osteoid, and nuclear pleomorphism. Usually, osteosarcomas arising in the appendicular skeleton have a high grade with respect to the cranial ones.

What WHO underlines for human species is important also in Veterinary medicine: because of the pitfalls and limitations of grading, some rules must be respected: grading is not a substitute of an accurate histological diagnosis; it requires representative and well-processed material that should be obtained before neoadjuvant therapy. Several criticisms have been made of histological grading. A universal grading system is not possible for all sarcomas, but it is unrealistic to develop a grading system for every specific histological type of sarcoma and the systems currently used to perform correctly for the most frequent sarcoma types and represent an acceptable alternative. The reproducibility of the same grading system among pathologists and of different grading systems for the same tumors is rather poor. Most grading systems are three-grade systems with an intermediate grade that actually corresponds to undetermined prognosis and represent about half of cases. The universal use of needle-core biopsies is another important limitation for grading. Although grading on needle-core biopsies has been reported to show an acceptable degree of accuracy, determination of grade can be done with certainty for high-grade sarcoma only. Histological grading may be helped by imaging procedures for evaluating necrosis and MIB1 score instead of mitotic index. Despite its limitations, grade remains the most important prognostic factor in most sarcomas, and it is clear that clinicians will continue to expect pathologists to provide a grade for most sarcomas.

Biomolecular Highlights: We just investigated several times with canine model in translational cancer researches to compared biological, biomolecular, and clinical characteristics of human and canine osteosarcomas to try to identify common etiopathogenetic mechanisms. In the canine and human species, osteosarcoma is the tumor of bone with the highest frequency, with a value of about 80–85% (in respect of all other bone tumors), a high degree of invasiveness, and a high rate of metastasis and malignancy. Humans and dogs have many genetic and biomolecular similarities such as alterations in the expression of p53 and in some types of microRNAs that our working group has already described previously in several separate works. In our last paper, published in Cells in 2021, we report and collect new comparative biomolecular features of osteosarcoma in dogs and humans, which may represent an innovative update on the biomolecular profile of this tumor (Fig. 2.34).

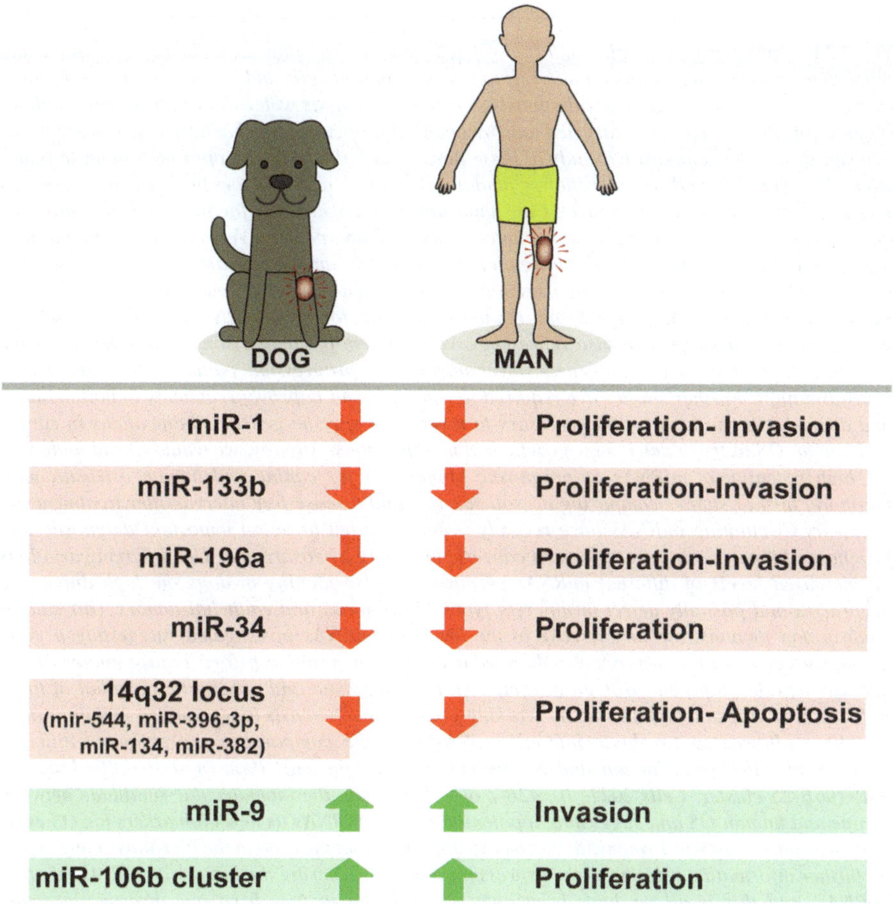

Fig. 2.34 Different levels of expression of tested microRNAs in biological behavior of canine and human osteosarcomas. Extract from: Leonardi et al. (2021)*; published open access, distributed under the terms and conditions of the Creative Commons Attribution (CC BY) license* (http://creativecommons.org/licenses/by/4.0/): *MiRNAs are highly conserved non-coding RNAs of ~ 22–25 oligonucleotides that exert a primary modulatory role on gene expression, regulating more than 60% of the genome post-transcriptionally. Over the years, biomolecular studies on the expression of some biomarkers, such as microRNAs, have allowed us to identify forms of altered regulation of the same type in different forms of spontaneous tumors; our studies have been established how these can play the role of tumor suppressor or tumor promoter in the co-involvement of specific targeting genes, resistance to apoptotic phenomena of cell death, the control of angiogenesis,* etc. *The correlation between these altered expressions of microRNAs has also allowed us to understand how these can play a fundamental role in the identification and application of new therapeutic models.Moreover, miRNA alterations correlate with a more aggressive behavior of cancer, making these molecules potentially prognostic and therapeutic targets also for osteosarcomas.In the last 20 years, comparative oncology has quickly expanded and the study of spontaneous cancers in the domestic dog has provided a suitable and interesting model for understanding several etiopathological mechanisms and, above all, new diagnostic tools and therapeutic paths for cancer in humans.We are confident that continuing on this template in the*

Fig. 2.34 (continued) *exploration and study of new markers will allow us to identify innovative and hitherto unknown aspects that will allow us to establish new genetic and biomolecular investigation methods for an increasingly early diagnosis of osteosarcoma, as well as to establish new methodologies for an increasingly targeted and decisive therapeutic approach. We are also firmly convinced that the comparative study of these parameters allows for an approach to these issues with a broader view and consideration of fundamental comparative factors between species, which recognizes the canine study model as an important reference factor for also understanding the oncological and osteosarcomatous dynamics of the human species. We well know how the new therapeutical approaches respect conventional treatments, and are clearly required, especially when involving biomolecular and personal pharmacogenomics therapies, using specific biomarkers and biomolecular targets, and the use of animals for comparative therapeutic considerations; these approaches can also represent a valid system of comparison and reference for the understanding of the etiopathogenetic mechanisms and progression paths of deep cell tumor transformation. Further studies are required to establish and improve treatments in both humans and dogs. Additionally, it is still necessary to investigate specific genetic modifications in canine and human OS to try to detect new genetic loci modifications. Anyway, we would like to underline the high potential of miRNAs as prognostic biomarkers for canine and human osteosarcoma; therefore, further studies comparing overall survival and disease-free interval after treatment are necessary.Circulating miRNAs have recently been suggested to be an important diagnostic and prognostic biomarker in several human cancers, including osteosarcoma. The different variations on increased levels of different miRNAs permit us to also identify in dogs the high diagnostic accuracies and partially detect in different types of sarcomas and epithelial tumors. Our studies confirm and demonstrate an increase in the levels of miRNAs investigated, suggesting a good diagnostic accuracy that was not well detected in comparative studies before. Future studies will be necessary to develop new strategies to improve the diagnostic and prognostic potential of these miRNAs in canine and human OS.Our working group already boasts long-standing collaborations that have allowed us to obtain interesting results on the comparative expression of different isoforms of miRNAs in human and canine OS, such as the miR-196a, miR-1, miR-133b, and miR-106B-25 cluster, Cells 2021, 10, 428 7 of 9 confirming the biomolecular similitude between canine and human OS and suggesting a potential role of miRNAs as new biomarkers for OS early diagnosis and treatment.From what can be extrapolated from the data in the literature, there is still no further information relating to the expression of factors altering the activity of miR-9 and other miRNAs, and this is all we have been able to obtain up to now from our investigations and recognitions in the related data banks. This lack of data, however, will quickly represent a further stimulus to continue on the path of comparative biomolecular evaluation on these and other factors, which to date still remain unexplored, as well as those of several exosomes, premetastatic niches, and other factors that we have set out to investigate together in future comparative works on human and canine osteosarcoma. Future studies of these factors and those of other biomolecular and genetic aspects—aimed at an increasingly sophisticated characterization of osteosarcomas—are still so devastating and lethal for many of their biological aspects and will certainly guarantee to be able to identify new innovative diagnostic and therapeutic paths, in order to guarantee the identification of new access "keys" for ever more early and precise diagnoses and for ever more targeted and decisive therapies. The animal model, also from a "One Health" perspective, will be able to ensure for a long time a study model on which to build new comparative safeguards for the fight against many diseases, including those of a more difficult approach, such as those of oncological type*

Central Osteosarcoma
(Conventional Osteosarcoma, Classic Osteosarcoma)

Definition: A high-grade malignant neoplasm with malignant osteoblasts producing osteoid.

Etiology: usually arise primary without well-known predisposing conditions. Described also in animals with other previous diseases of bone as like radiation exposure, traumatic injuries, and foreign bodies comprising orthopedic implants, genetic abnormalities, chemotherapy, described in association with parasitic infestations such as that of *Spirocerca lupi* and also in other conditions with a specific genetic susceptibility as retinoblastoma with germline mutation for RB gene and P53.

Image Findings: frequently plain radiography can be fundamental and helpful for a better diagnostic setting before obtaining the diagnosis of histopathological certainty. Osteosarcoma has destructive and infiltrative lesions, frequently localized around the metaphyseal areas of long bones. Imaging show a mixed pattern of radiolucency and radiodensity of involved bones with different stages of gravity. The lesions usually involve cortex, with periosteal reaction characterized by reactive bone between cortex and periosteum with an appearance of "sunburst" when radiating and when in multiple layers of "onion skin." The typical appearance of periosteal reaction localized at diaphysis involved cortex and periosteum is calling "Codman triangle."

Macroscopic features: most frequently arises from the metaphyseal region, but can originate from any bone portion, as a large mass. Tumor is usually gritty, tan-gray in color and with amounts of mineralized bone. The tumor may also contain cartilage, cysts, hemorrhages, mucinous material, necrosis in relation to its main histological characteristics related to the subtype. The neoplastic growth may involve periosteum, cortex, medullary cavity with evidence of layer of reactive bone, involvement and destruction of periosteum, and total invasion through the physis of cortical surface, to articular surface synovial membrane, joint capsule, and adjacent soft tissues. Skip metastases in the same affected bone are frequent involving medullary cavity far or main to the primitive mass. Anyway, one of the most diagnostic features of central osteosarcoma is the massive invasion of the marrow space with strong reaction from the host characterized by inflammation atrophy of neighboring soft tissues.

Microscopic pathology: The microscopic appearance of conventional osteosarcoma has a wide range of cellular morphology related to the histotype prevalence. The general features are characterized by the population of neoplastic cells where is possible to detect also other elements in different proportion and represented by spindle-fusiform cells, ovoid cells, plasmacytoid and epithelioid cells, multinucleated osteoclast-like giant cells. Always there is production of bone matrix produced by malignant cells. The

(continued)

atypia grade is always high and the neoplastic cells show cytoplasms vary in size and always eosinophilic. Nuclei are in central or eccentric position, with strong hyperchromasia and prominent nucleoli. Mitotic atypical figures are very frequent.

Main histologic types differences:

POORLY DIFFERENTIATED OSTEOSARCOMA consisting mainly of mesenchymal cells with high grade of anaplasia and pleomorphism.

OSTEOBLASTIC OSTEOSARCOMA is characterized by abundant but variable amounts of deposition of osteoid (productive and nonproductive form).

CHONDROBLASTIC OSTEOSARCOMA there is a predominant cellular component of cartilage.

FIBROBLASTIC in this subtype spindle cells predominates.

TELANGIECTATIC OSTEOSARCOMA is a very aggressive sarcoma characterized by blood-filled cysts lined by tumor cells and not by endothelium as in hemangiosarcoma. In this subtype there is also production of osteoid.

GIANT CELL OSTEOSARCOMA Osteoclast-like giant cells predominate in this form.

Peripheral Osteosarcomas (Surface Osteosarcoma, Juxtacortical Osteosarcoma)

This separate classification includes osteosarcomas arising from the surface of bone near the periosteum. Juxtacortical osteosarcoma is the term that includes a less common group of osteosarcomas having lower grade features of malignancy and biological behavior as compared to osteosarcomas that arise in the medullary cavity. Examples of this group in order of highest to lowest grade of malignant behavior include: high-grade surface osteosarcoma (a tumor while infrequently diagnosed in dogs is not yet included in the WHO Classification for Animal Bone Tumors), periosteal osteosarcoma, and parosteal osteosarcoma.

2.12.3.2 Periosteal Osteosarcoma

Malignant bone tumor of intermediate grade arising in periosteum generally of long bones with predominant chondrosarcomatous components that tend to infiltrate also the cortex from the external surface. In humans it accounts for 1–2% of all osteosarcomas and the most common bones involved from this subtype are mid or distal femur diaphysis and proximal tibial diaphysis followed by ulna and humerus.

Usually consists in a small to moderate size of well-demarcated firm palpable mass with mild to moderate pain and slow or rapid growth. The margins are well circumscribed with broad base anchored to underlying cortex that may be invaded to the reach the medullary cavity. Periosteal osteosarcoma can have a consistency varying from soft to rubbery with abundant chondroid components that give an additional feature of translucent bluish tissue without clear appearance of well-differentiated hyaline cartilage. In the deep layers the tumor may be gritty and with a well-differentiated bone texture.

Cytologically periosteal osteosarcoma in fine needle aspirates smears can present differentiated and variable aspects with low or moderate cellularity to others with higher cellularity mixed with a significant background of cellular elements of the chondroid or chondro-myxoid type. Osteosarcoma's cells are round to oval in shape with moderate sometimes vacuolated cytoplasm resembling an appearance of chondrosarcoma. With higher malignant cellular characters, the pleomorphism of neoplastic cells is always present as like as anisocytosis and anisokaryosis with hyperchromatic nuclei and coarse chromatin. In these cases, atypical mitoses are always present as like as binucleate and multinucleated cells. In periosteal osteosarcoma smears are possible to find frequently fragments of hyaline cartilage.

These tumor cells seem to arise from the periosteum, from cellular precursors differentiating in chondroblasts and osteoblasts.

Histological features of periosteal osteosarcoma are represented by multiple and irregular lobules of predominant hyaline cartilage and bone, oriented at around 90 degrees to the cortex layer. Usually, these tumors are not well differentiated and the cells should be confused with reactive bone cells, but chondrocytes may show different grade of atypia and mitotic figures. These malignant cells are always mixed with other atypical cells with spindle or polymorphic shape.

In human periosteal osteosarcoma it is considered a tumor with a lower malignancy index than central osteosarcoma. There are very few reports of periosteal osteosarcoma in domestic animals and these include the canine and equine species where forms of different malignancy have been reported. Obviously, even the diagnostic identification of periosteal osteosarcoma requires an early clinical identification of the tumor before its growth involves entire segments and bone structures, causing a loss of primitive onset and expansion references. The differential diagnosis in domestic animals is related to all the other neoplastic diseases characterized with the presence of cartilage as like as osteochondroma, chondroma and chondrosarcoma, parosteal osteosarcoma, and chondroblastic osteosarcoma (Figs. 2.35, 2.36, and 2.37).

Periosteal Osteosarcoma

Definition: Malignant bone tumor of intermediate grade arising in periosteum.

Etiology: mutation in mesenchymal stem cells.

Image Findings: mass with broad base with reactive bone generally radiating from it. Cartilaginous detectable tissue and not frequent invasion of underlying bone.

Macroscopic features: Solid, whitish-cartilaginous mass with different elongated shapes and well-circumscribed margins.

Microscopic pathology: typical pattern of sarcoma with cartilaginous (matrix) and new bone components.

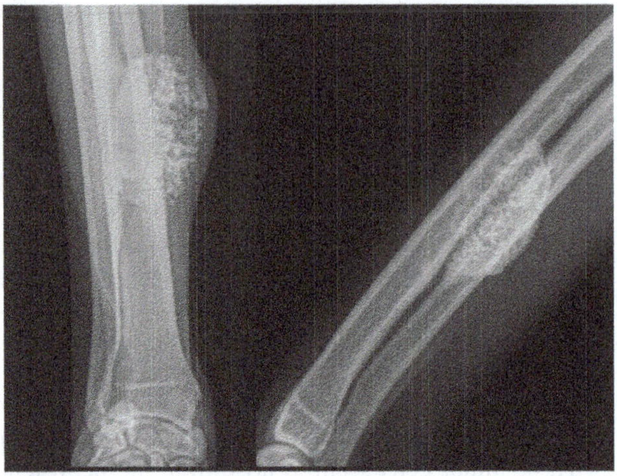

Fig. 2.35 Periosteal osteosarcoma in a 3-year-old NM Labrador-mixed dog that presented with a nonpainful hard nodular enlargement on the dorsomedial surface of the distal radius. Periosteal mass has contoured surface, does not appear to invade the surface of its cortical base, and is composed of multiple ossified densities separated by soft tissue (Courtesy of Dr. Roy R. Pool, Emeritus Prof. UCD and Texas A&M University)

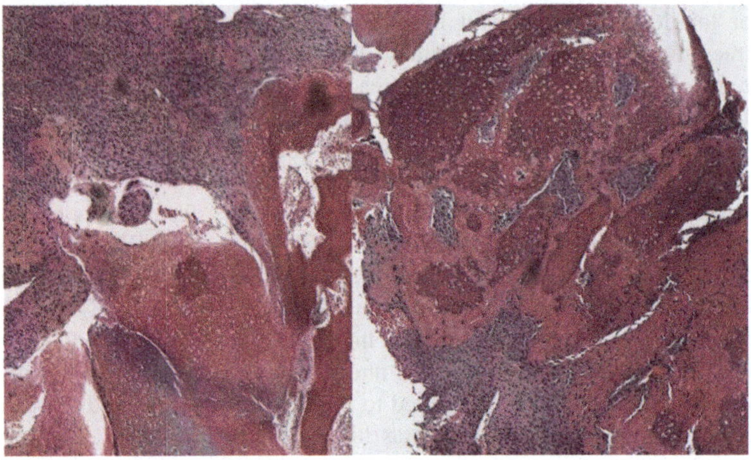

Fig. 2.36 Periosteal osteosarcoma: Tumor mass of small hyperchromatic cells has formed and entrapped themselves in random nodules of immature bone. H&E stain (Courtesy of Dr. Roy R. Pool, Emeritus Prof. UCD and Texas A&M University)

Parosteal Osteosarcoma

This is also a surface form of osteosarcoma (also called juxtacortical or surface osteosarcoma) represented by well-differentiated and intermediate grade of malignancy superficial osteosarcoma of the periosteum is only recently being diagnosed in increasing frequency in domestic animals, more often on a bone surface of the head than an appendicular bone of dogs.

Due to the scarce data in the veterinary literature, we must necessarily refer in a comparative way to those reported in human medicine where parosteal osteosarcoma is frequent as surface low-grade primary tumor (65% of all surface osteosarcomas)

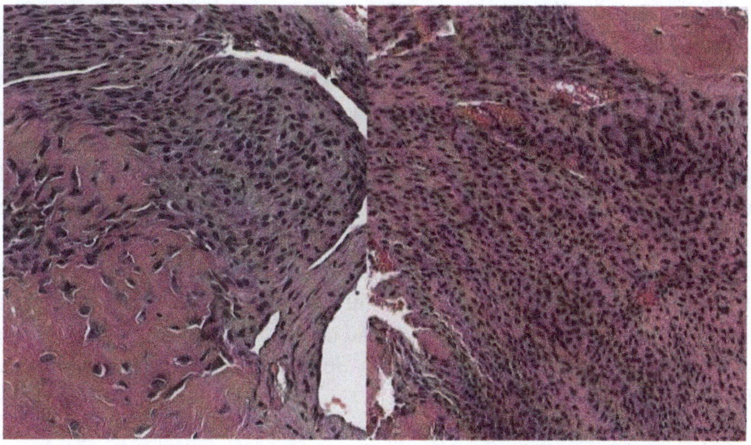

Fig. 2.37 Periosteal osteosarcoma: Small hyperchromatic spindle cells with angular profiles but without anaplastic features or mitotic figures form and entrap themselves in immature bone matrix. H&E stain (Courtesy of Dr. Roy R. Pool, Roy R Pool, Emeritus Prof. UCD and Texas A&M University)

with better prognosis with respect to central conventional osteosarcoma and accounts for 5% of all osteosarcomas. For the frequent modalities of growth, often and initially asymptomatic, parosteal osteosarcoma is frequently detected when the tumor becomes big in size.

In man parosteal osteosarcoma arises on tubular bones with most common site represented by femoral cortex while it has very uncommon locations in spine, pelvis, or other bones as craniofacial ones. Parosteal osteosarcoma arises as a firm, palpable, slow enlarging, not always painful mass with a smooth surface and fibrous capsule. Image findings are characterized by an exophytic mass, multilobulated in most cases, uniformly well mineralized in the cortical surface (intact differently from periosteal osteosarcoma where involvement and cortical invasion occur fast) with broad sessile or pedunculated base, sometimes thickened cortex but with little or in absence of periosteal reaction. When parosteal osteosarcoma extends to cortex and medullary cavity this pattern can be well detected with CT and MRI. Angiography can help too to identify areas of hypervascularity important to identify better more aggressive or more atypical forms of the tumor.

Tumor matrix may show different patterns of density as result in radiographic investigations with maximal radiodensity near the implant area of the tumor. Radiodensity may be produced by the irregular production of a net of trabeculae of bone that give also the radiological characteristic typical steel-wool pattern.

Histologically it is frequently not so easy to make a histological diagnosis of parosteal osteosarcoma without a support of radiographic images.

In Veterinary Medicine this tumor was described in cats with few cases in literature as like as in canine species.

Histological features of parosteal osteosarcomas are usually represented by well-formed osseous trabeculae of woven bone not rimmed by osteoblasts embedded in a mixed of spindle cells and well-differentiated collagen fibers most of these of collagen and sometimes mixed with focal island of cartilage. In rare cases was described also the presence of a cartilage cap that can necessarily involve in differential diagnosis cases of osteochondroma. Mitotic index is frequently low, but due to the continuous growth of the tumor parosteal osteosarcoma becomes more aggressive and with invasion of cortex and medullar cavity it can acquire a higher index of malignancy with signs of cellular atypia more similar to what happens in the previously described forms of conventional osteosarcoma, with the appearance of atypical mitotic figures and metastatic spread of the tumor to adjacent bone segments and to the lungs as highlighted and also described in some reports of cases in animals.

Stromal cells contain elongated nuclei and low to moderate signs of cellular atypia and sometimes resembling fibromatous cells with pale eosinophilic cytoplasm. Components of the tumor with high grade of cellular dedifferentiation contain nuclei with coarse chromatin and mitotic activity as like as appear also in malignant cartilaginous cells (chondrosarcomatous foci mimic osteochondroma where, however, there is no presence of spindle components) (Fig. 2.38).

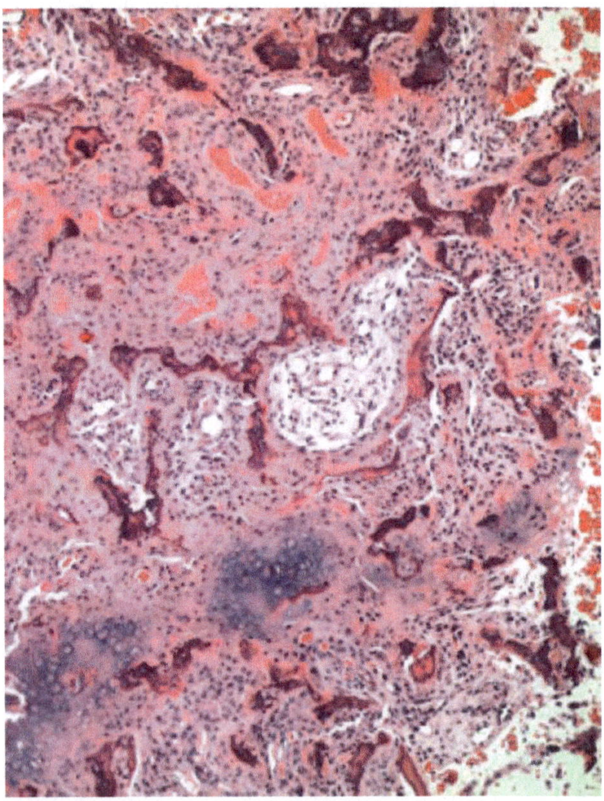

Fig. 2.38 Parosteal osteosarcoma from humerus of a dog (Courtesy of Dr. Roy R. Pool, DVM, PhD, Emeritus Professor of Pathology UCVD and Texas A&M—USA)

Parosteal Osteosarcoma

Definition: surface form of osteosarcoma (also called juxtacortical or surface osteosarcoma) represented by well-differentiated and intermediate grade of malignancy superficial osteosarcoma originating from the external fibrous part of periosteum very rarely described in animals.

Etiology: unknown.

Image findings: Well-circumscribed mass mineralized uniformly without periosteal reaction with no or low invasion of underlying bone.

Macroscopic features: Solid, hard, with tan areas corresponding to proliferation of fibroblastic parts.

Microscopic pathology: presence of well-differentiated woven bone trabeculae surrounded by a layer of spindle atypical pleomorphic cells and collagen (Fig. 2.39).

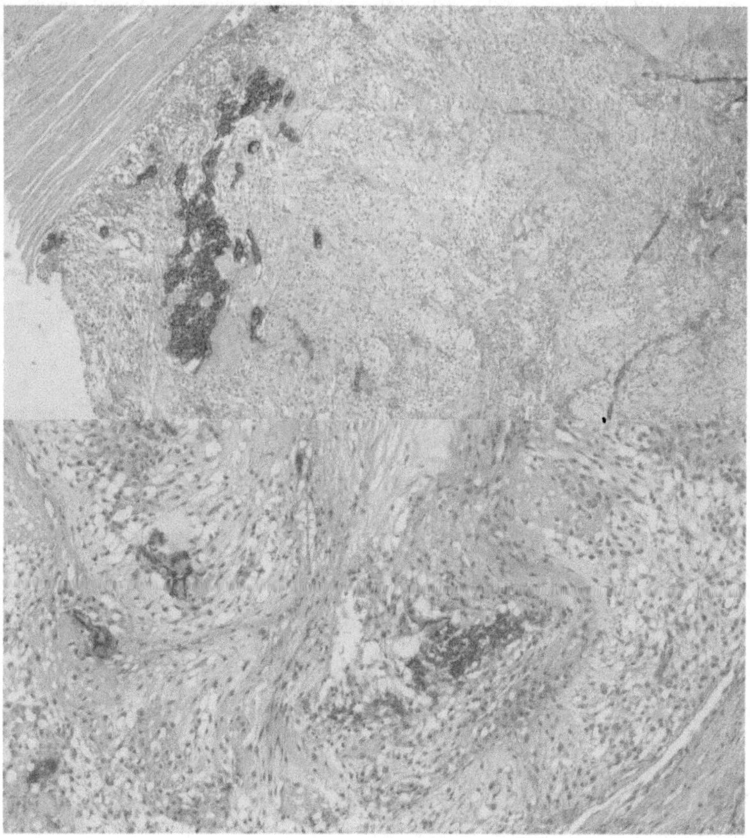

Fig. 2.39 Dog, Golden Retriever, adult, female. "Mixed" osteosarcoma with rare myxoid-fibrillar pattern and a low mitotic index. For histomorphological aspects this type of osteosarcoma could be comparable with a form of human osteosarcoma, with a low grade of malignancy, called chondromyxoid fibroma-like osteosarcoma (CMF-OS) an exceedingly rare subtype of central osteosarcoma, not classified in veterinary medicine (from author collection)

Extraskeletal Osteosarcoma

Extraskeletal osteosarcoma represents a primitive osteosarcoma arising in tissues other than bone without primary involvement of the skeletal system. Extraskeletal osteosarcomas have been described in several animal species, with dogs and cats being the most affected with a higher frequency in mammary glands followed by other organs and tissues as spleen, subcutaneous tissue, gastrointestinal tract, skin, muscle, liver, urinary system, thyroid glands, salivary glands, without predilection of specific breeds.

Usually, extraskeletal osteosarcoma consists in an infiltrating mass with rapid growth with irregular internal areas of ossifications mixed with hemorrhage and necrosis. In all tissues, especially in mammary glands, it can arise from metaplastic bone foci or from malignant transformation of myoepithelial cells.

Extraskeletal osteosarcoma has histological features similar to the other osteosarcoma of bone and is always highly malignant and metastasizing. In dogs was reported a frequency of metastases of 64% with 7 cases to 11 of extraskeletal osteosarcoma with involvement of lymph nodes, liver, lungs, and stomach. In human medicine Campanacci described the frequent presence of malignant cartilage without evidence of dominant feature of chondroblastic type and with rare case of well differentiated and low grade of malignancy of extraskeletal osteosarcomas. The specific treatment provides surgery (wide or radical) and chemotherapy similar for osteosarcoma primitive of bone (Fig. 2.40).

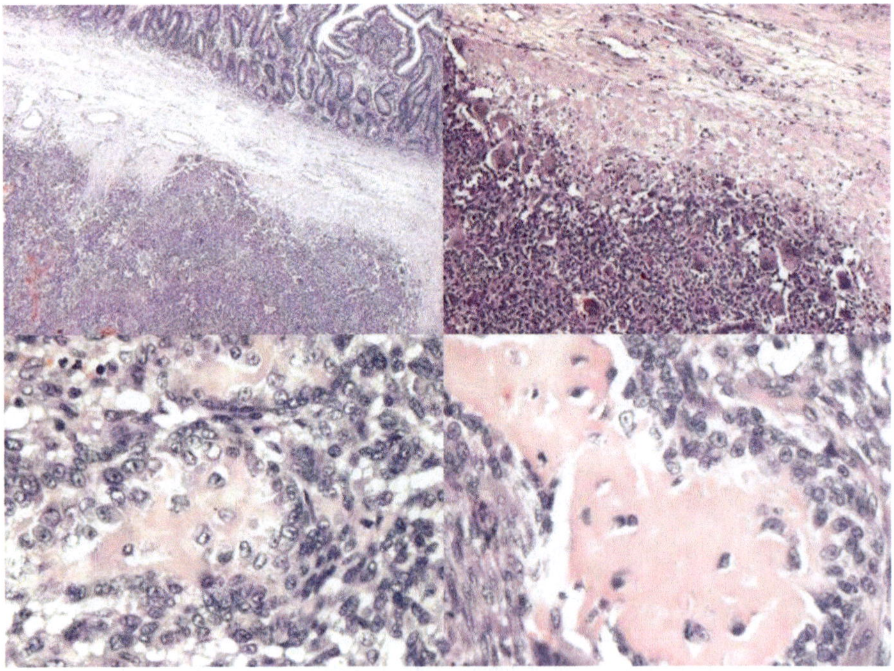

Fig. 2.40 Extraskeletal intestinal osteosarcoma in a 13-year-old male, Cocker Spaniel dog. H&E stain (Leonardi et al. 2012b)

2.12.4 Chondrosarcoma

Chondrosarcoma is a primary malignant bone tumor producing cartilaginous and fibrillar matrix without osteoid but which can contain bone usually originating by endochondral ossification from malignant cartilage and malignant mesenchymal cells. Chondrosarcoma originates often centrally in the bone (central chondrosarcoma or intramedullary), or peripherally from the surface of the bone (also called juxtacortical). Usually it is a slow-growing tumor detected frequently with few symptoms as local swelling and pain. In aggressive malignant metastatic forms chondrosarcomas tend to invade blood vessels with late appearances of metastasization to the lungs and very rare involvement of lymph nodes. After surgical excision these tumors may have also a high frequency of recurrence.

In animals the most frequent forms are represented by central chondrosarcoma, where cats and dogs are the more representative affected species, while in sheep it represents the most frequent bone tumor. Chondrosarcomas are the second bone tumor for frequency in domestic animals and usually have fewer general characters of malignancy and metastasization with respect to the most common osteosarcoma. Primary chondrosarcoma arises in normal bone or rarely also in extraskeletal soft tissues and secondary chondrosarcoma can arise in association and after malignant changes of other pre-existing neoplastic or non-neoplastic injuries as osteochondroma, enchondroma, fibrous dysplasia or exostosis or, in other hand, also in tissues previously treated with radiations. Diaz-Betrana C. et al. and Aeffner F. et al. in two separate papers described in dogs malignant transformation of synovial chondromatosis to chondrosarcoma.

Dogs represent the species where chondrosarcoma is most frequent with a 10% of frequency with respect to all bone tumors secondary only to osteosarcoma. Strangely in cat species where cartilaginous disorders are several and relatively frequent chondrosarcoma is not common and detected with a low frequency.

Few epidemiological data in literature report a higher incidence in large canine breeds, rarer frequency in small breeds and also in giant breeds where osteosarcoma is, however, always frequent. Amy Durham et al. described with a retrospective study the feline chondrosarcomas in 67 cats with a mean age of 9.6 years in a range from 2 to 18 years of age where males were affected more often than females.

Radiographic features of chondrosarcoma show frequent and clear evidence of malignancy with typical moth eaten osteolysis and multilobulate patterns of cartilage. The growth of the tumor involves the epiphysis through the epiphyseal cartilage. Chondrosarcoma invades also the cortex and reaches periosteum where it produces an intense periosteal reaction forming of osteo-cartilaginous malignant spicules. Chondrosarcoma usually arises in bones originating from endochondral ossification and in long bones it involves usually primarily metaphysis or/and diaphysis, but osteochondromas may affect also small bones and skull. Pathologic fractures are not frequent. Occasionally chondrosarcoma can arise in no skeletal soft tissues and it is called extraskeletal chondrosarcoma.

Clinical presentation of chondrosarcoma is characterized by enlarging mass, pain as a constant and primary symptom. The growth is usually slow but there are also cases with rapid growth and aggressive behavior. The tumor is often large but variable in size, generally multilobular, with typical area of blue-gray-whitish central osteolysis, permeative, and the presence of malignant cartilage may vary with the tumor grade. Due to its potential of undifferentiation, it is important to underline that, especially in the poorly differentiated forms, the tumor mass can assume variable aspects similar to those osteosarcomatous or with variable patterns also defined as brain-like or plump, but, in any case, chondrosarcoma always maintains general characteristics of evident presence of malignant cartilage and that the gross appearance is conditioned by own components as hyaline-myxoid cartilage, invasiveness, presence of chondroid matrix and/or mineralization with calcification-ossification of the cartilage. Chondrosarcoma may appear grossly very similar with chondroma especially when composed of well-differentiated hyaline cartilage with a typical bluish color, soft, jelly, and translucent. Usually, these tumors are avascular with rare undifferentiated cases that may show hemorrhagic and soft, white-yellow necrotic peri lobular foci frequently mixed with white areas of calcification.

Generally, the mass has a smooth surface but often when the osseous cortex is thickened chondrosarcoma may have a rough surface due to its permeation in the systems of Haversian canals that activate a chronic periostitis with also production of different layers of reactive bone and irregular mineralization of the surface of the bone. Sometimes chondrosarcoma invades the joint space. Anyway, due to its multiple forms with which chondrosarcoma can occur it is practically impossible to differentiate this neoplasia from other tumors with a similar radiographic, clinical, and macroscopic appearance such as osteosarcoma or chondroma.

A particular aspect of canine chondrosarcoma of the rib is the location close to osteochondral junction where cartilage is strongly present. Other sites where chondrosarcoma can arise are represented by nasal cavity, vertebrae, scapula, etc., all of these less frequent with respect to long bones and extra skeletal locations where also chondrosarcoma may acquire more aggressive behavior.

CT scans may be useful to identify clearly matrix calcifications instead MRI is more helpful to evaluate the extent of the tumor and the potential involvement of surrounded soft tissues.

Interesting data from human species describe some forms of primary central chondrosarcomas founded in association with breast cancer, positive for estrogen receptors in early age.

Cytological features of chondrosarcoma are strictly related to the grade of differentiation and malignancy of the tumor but the constant feature is the presence of chondroid matrix and cartilaginous lacunae (containing chondrocytes), mixed with malignant spindle or pleomorphic cells, similar to osteosarcoma or undifferentiated sarcoma. The cytological cellular population similar to the osteosarcomatous condition is characterized by low cellular aspirates containing malignant cells with different shapes from round to oval or also spindle-fusiform with large nuclei, hyperchromatic in a basophilic, and frequently scant to moderate amounts of vacuolated cytoplasm with red granules. In less malignant chondrosarcoma the single chondrocytes show enlarged nuclei with coarse chromatin and prominent nucleoli in absence of mitotic figures and with rare presence of binucleate cells. In aggressive malignant other cases of chondrosarcoma, the smears appear with moderate cellularity, a marked presence of different amount of myxoid cartilaginous substance with a fibrillar appearance. Chondrocytes inside the lacunar spaces may have two or more large nuclei in a general plasmacytoid shape and with cytoplasmic vacuoles, with coarse chromatin and one or more nucleoli. For histopathological evaluation would also be suggested using Romanowsky and Papanicolaou stains, in addition to the classic hematoxylin-eosin, which allow a better visualization and a good evaluation of the single chondrocytes and their general morphological details.

Histological features of chondrosarcoma have a broad range of aspects relating to the grade of differentiation and the cellular alterations and malignancy index that can vary from low-grade lesions (like in chondroma) to severe high grade more like to chondrosarcomas. Probably chondrosarcoma represents the tumor where differentiation it from the related benign form as chondroma it is one of the most complex evaluative diagnostic steps in veterinary pathology also because the benign form may undergo a progressive transformation to malignancy and the same tumor may show contemporarily different pattern of cellular modification. Small core biopsies are not always representative also in well-differentiated lesions. The main diagnostic problems are be able to distinguish benign forms as chondroma or synovial sarcomatosis, from malignant form of different grade of malignancy as like as also chondroblastic osteosarcoma. For these reasons is very important to try to evaluate all cases from different views: clinical and imaging findings may be very helpful to evaluate macroscopic characters of the lesions and also histological investigations that give morphological and staging of the tumor. To obtain a diagnostic biopsy in case of chondrosarcoma is always a problem, especially when the tumor is central and where the presence of eosinophilic hyaline cartilaginous matrix should be very similar to osteoid and lead to evaluation errors that can lead to a misdiagnosis as like as when chondroid matrix is changing for endochondral ossification.

Histologically chondrosarcoma consists of multilobular mass composed of lobules of malignant mesenchymal cells that can produce different amounts of blue-gray chondroid mucinous matrix (myxoid cartilage) similar to hyaline cartilage with the potential presence also of mineralization, ossification, hemorrhages, and necrosis. Frequently the lobules are surrounded by variable amounts of spindle cells arranged in fibrous bands and inside the lobular structures is possible to detect malignant chondrocytes arranged in diffuse agglomerates and may permeate bony trabeculae. Malignant features in chondrosarcoma are represented by high cellularity and significative presence of cartilage cells, varying in shape and size, frequently referring to malignant cells originating from hyaline cartilage or others precursor of cartilage not excluded embryonic cartilage, and permeation of cortical and/or medullary bone. These cells have different grades of anisocytosis and anisokaryosis with enlarged hyperchromatic nuclei, frequent binucleate cells, and medium-high mitotic index, also if some authors suggest that at least also one mitotic figure in chondrosarcoma may orient the pathologist to diagnosis of chondrosarcoma.

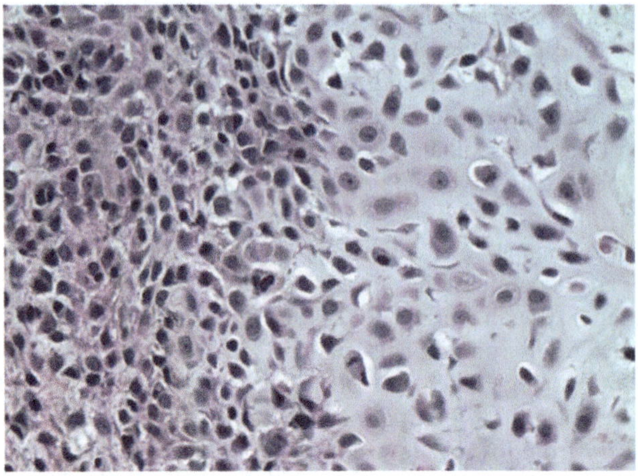

Fig. 2.41 Chondrosarcoma from femur of a dog (Courtesy of Dr. Roy R. Pool, DVM, PhD, Emeritus Professor of Pathology UCVD and Texas A&M—USA)

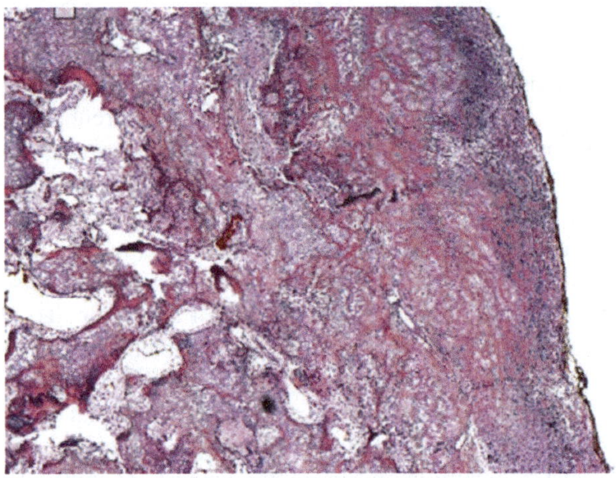

Fig. 2.42 Perichondrial sarcoma, i.e. chondrosarcoma of periosteum from shaft of radius of a dog (Courtesy of Dr. Roy R. Pool, DVM, PhD, Emeritus Professor of Pathology UCVD and Texas A&M—USA)

Changes of chondroid matrix in myxoid tissue or with a fluidification of chondroid matrix are common in chondrosarcoma.

In humans grading of chondrosarcoma is very important to predict the histological behavior and prognosis and the authors would like to propose it as like as classified by WHO in human conventional chondrosarcomas with 3 different groups: I, II, III. The grading is based first on cellularity, nuclear stain attitude that characterize hyperchromasia, nuclear size, and mitotic index (Figs. 2.41 and 2.42) (Table 2.4).

In dogs and in extra nasal chondrosarcomas histological grade is prognostic as reported also by Waltman S.S. a paper in 2007 about a study in 25 dogs affected and amputated with diagnosis of appendicular chondrosarcoma.

Immunohistochemistry in well-differentiated chondrosarcoma stain cytoplasmic positivity for S-100 protein for cartilaginous cells, while the immunoreactive in

Table 2.4 Cytological aspects of different grade of malignancy in chondrosarcoma

Low grade (I)	Intermediate grade (II)	High grade (III)
Low cellularity	Low to moderate cellularity	High cellularity
Particles of hyaline cartilage with lacunar spaces containing well-differentiated chondrocytes	Evidence of important amounts of myxoid chondroid matrix with atypical spindle cells may be present.	Abundant myxoid chondroid matrix
Chondrocytes round to plasmacytoid and occasionally binucleated	Single malignant cells or in small clusters internal to chondroid lacunae	Marked cellular pleomorphism
Cytoplasm vacuolated. Possible to detect cells with signet ring pattern	Moderate to abundant vacuolated cytoplasm	Chondrocytes single or in small clusters and peripheral lobular cells are less differentiated and more spindle in shape
Nuclei of malignant cells enlarged with granular chromatin and one or more distinct nucleoli, hyperchromatic, large and uniform in size.	Nuclei appear hyperchromatic, lightly enlarged with one-tow nucleoli and granular chromatin	Atypical nuclei with coarse chromatin and one or more nucleoli in very hyperchromatic cellular background
Absence of mitotic figures	Mitoses can be found	Mitotic figures may be present
Few amount or absence of chondroid-myxoid matrix.	Chondroid-myxoid matrix more evident than hyaline normal cartilage.	Chondroid matrix is always present with single free chondrocytes in the background or in lacunar spaces inside the chondro-myxoid matrix.

undifferentiated forms is weak. Chondrosarcomas are also vimentin positive and frequently when undifferentiated clear cells are present in chondrosarcomas may show cytokeratin and epithelial membrane antigen immunopositivity. Chondrosarcoma may also stain positive for osteonectin. Cellular proliferation activity may be detected with Ki67 generally increased in all grades of chondrosarcomas. To differentiate benign to malignant forms of chondrosarcoma may be useful also evaluate the immunohistochemical expression of different Metalloproteinases, always positive in malignancy also for other several forms of tumors. C-erbB-2 also known as *neu* or HER-2, a proto-oncogene of 190 kilodalton representing a transmembrane glycoprotein similar to epidermal growth factor (EGFR) is present in chondrosarcoma and is absent in all other benign cartilaginous tumors. Immunophenotype of human chondrosarcoma only in small percentage shows mutation of IDH1 (Isocitrate dehydrogenase 1 [NADP+]) that can be detected by the use of IDH1 R132H antibody to detect related mutation of the IDH1 R 132H as important marker of survival also in other type of tumors as in nervous and breast cancers (Table 2.5).

Table 2.5 In humans around 50% of isolated chondrosarcomas and enchondromas have mutations of IDH1 e IDH2. Enchondroma with IDH1 and IDH2 mutations as like as osteochondromas with EXT1 inactivation are composed of intermingled and wildtype cells (intraneoplastic mosaicism). For osteochondroma, it has been shown that the wildtype cells are prone to progress to malignancy since the majority of peripheral CS are EXT wildtype. Malignant transformation is caused by as yet unknown (epi)genetic alterations. Progression to high-grade chondrosarcoma is characterized by aneuploidy, defective cell cycle regulation, and activity of several signaling pathways as shown (mut, mutated; wt, wildtype). (From WHO classification of tumors of soft tissue and bone, 2013)

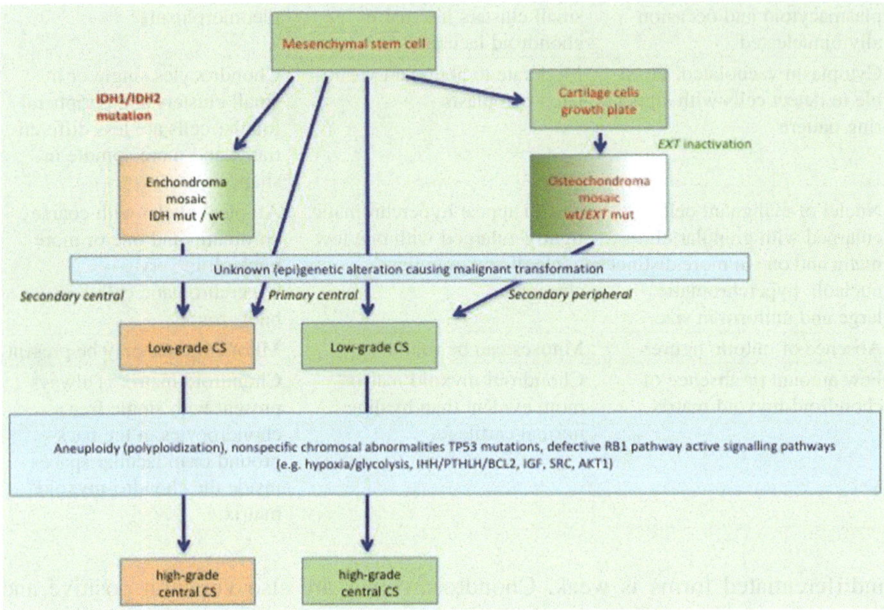

Chondrosarcoma

Definition: Chondrosarcoma is a primary malignant bone tumor producing cartilaginous and fibrillar matrix without osteoid.

Etiology: predominantly unknown, but in some cases is possible a malignant transformation from benign cartilaginous pre-existent tumors.

Clinical presentations: enlarging mass with pain and pathologic fracture when bone strongly damaged by the malignant growth.

Image findings: Lytic lesion with irregular bone density and mineralization of the surface or of the cortex.

Macroscopic features: Gray-tan color due to presence of hyaline cartilage mixed with translucent gray myxoid matrix. Possible find gritty areas referring to calcification and/or ossifications. The expansion of the tumor may involve

(continued)

medullary canal, cortex, periosteum, separately or together (medullary, central superficial forms).

Microscopic pathology: infiltrative multilobular tumor with fibrocartilage, myxoid tissue, homogeneous basophilic hyaline matrix, chondrocytes varying in size and shape with eosinophilic frequent vacuolated cytoplasms and different degrees of atypia. Hyaline cartilage composed by tumor round to oval cells located to lacunar spaces while in the myxoid areas the malignant cells may have a stellate bipolar shape and may are arranged in cords or single.

Differential diagnosis: Osteochondroma, fracture callus (smaller nodules of cartilage, cells clearly well differentiated, cartilage undergoes endochondral ossification).

Don't forget that: in most cases with ONLY biopsy it is almost impossible to interpret the differences between a low-grade chondrosarcoma from other benign cartilaginous tumor and that is ALWAYS advisable to provide a collegial evaluation to simultaneously interpret the characteristics of the clinical, radiographic, and anatomo-histopathological aspects of the tumor and its characteristics of infiltration and growth pattern (diagnostic).

2.12.5 Fibrosarcoma

Fibrosarcoma is a high-grade pure fibroblastic (spindle cell) non-osteogenic sarcoma of bones with different grades of malignancy and with tumor cells production of collagenous matrix and with typical organization in fascicular pattern (herringbone). It constitutes 5–9% of all primary malignant bone lesions. Fibrosarcoma arising frequently in medullary primary localization and in this case called also central or medullary fibrosarcoma, but there are also reports about periosteal fibrosarcoma of bone. Human pathologists recognize for fibrosarcoma synonymous names such as fibroxanthoma, malignant fibrous histiocytoma, malignant fibrous xanthoma, or undifferentiated pleomorphic sarcoma, but in veterinary medicine the name is recognized constantly in fibrosarcoma. Because of the potential for primary origin from different tissues, I would softly suggest to name of the primary skeletal/bone fibrosarcoma as "fibrosarcoma of bone" to better understanding of the primary origin of the tumor. Is always important to consider fibrosarcoma in differential diagnosis with fibroblastic osteosarcoma, poor differentiated osteosarcoma.

Fibrosarcoma of bone is rare in all species and represents a tumor of mature-old animals. Fibrosarcomas of bone are described mostly in dogs and cats, but also in horses and cows.

Meuten report a frequency of 5–9% with respect to all bone tumors with equal sex distribution.

There are not recognized predisposing factors in animals, but in humans about 20% of fibrosarcoma of bones are called secondary fibrosarcoma because growth in pre-existing bone lesions or also in association with prosthetic devices. In humans and in dogs was demonstrated the onset of fibrosarcomas of bone in tissue areas previously treated with radiotherapy protocols especially in cases of oral lesions.

Marconato and Del Piero report that bone fibrosarcoma mainly affects adult dogs of medium-large breeds, with an average age of about 8 years.

There are little informations about fibrosarcoma of bone in humans where genetic causes with large number of genomic imbalances have been detected in tumors with fibromatous differentiation.

In animals the presence of few cases described do not permit to identify site where fibrosarcoma rises most frequent. Long bones are the most affected sites in dogs where fibrosarcoma of bone was described in femur, tibia, radius, and ulna but it was describes also in less percentage in maxilla, mandible, pelvis, ribs, and vertebra. Usually arise in the area close to the edge from diaphysis and metaphysis of long bones (Fig. 2.48). In cats the tumor is also rare but was reported in olecranon, carpus, femur, and tibia. Fibrosarcoma has been reported also in nasal cavity, paranasal sinuses, and nasopharynx. Appendicular fibrosarcomas are less malignant with respect to oral localization and after amputation the prognosis is usually good also due to low metastatic rate (Figs. 2.43, 2.44, and 2.45).

Clinical symptoms are not specific and it is strongly related to the stage of the tumor, where it may vary from mild pain or indolent course (low grade) with very small mass to severe pain, with soft firm fast growth swelling mass, limitation of motion, and frequent pathologic fracture in cases of high grade of malignancy.

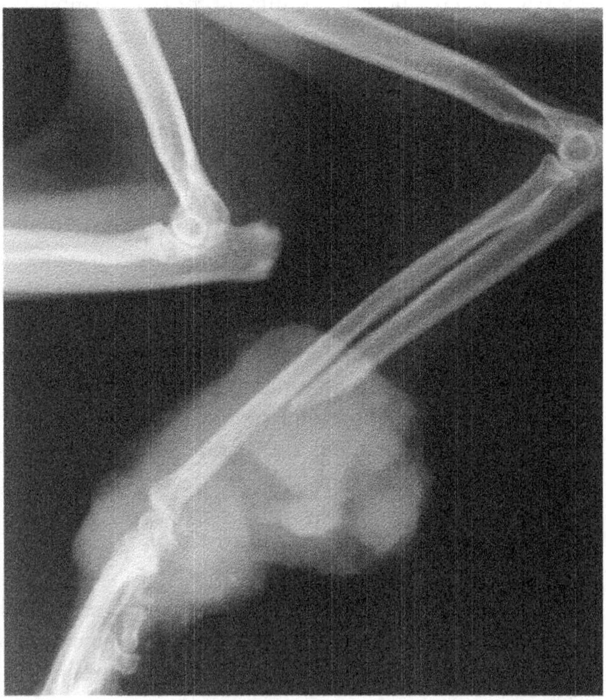

Fig. 2.43 Fibrosarcoma in a 5-year-old female, european cat ("Grigia"). Radiograph shows an extensive osteolitic-destructive mass involving metacarpus, carpus, radius, and ulna

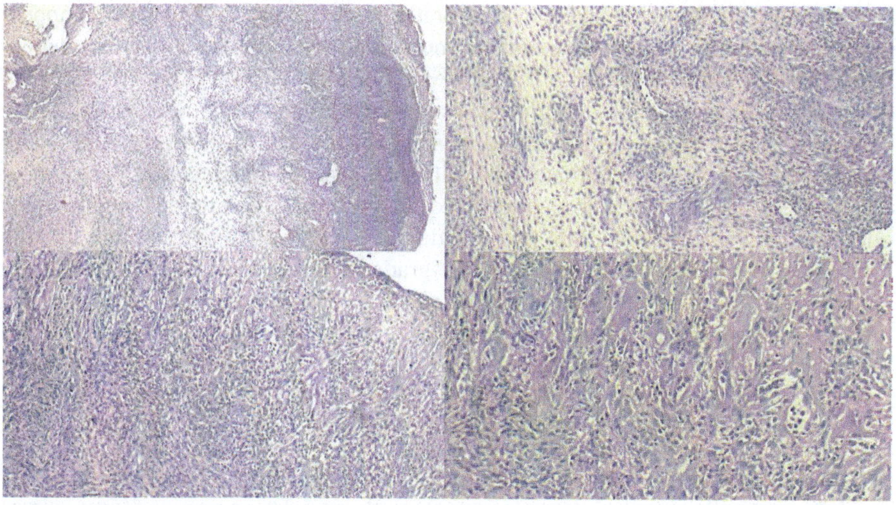

Fig. 2.44 Dog, maxillary fibrosarcoma. Typical feature of the tumor at several magnifications. (H&E, from author collection)

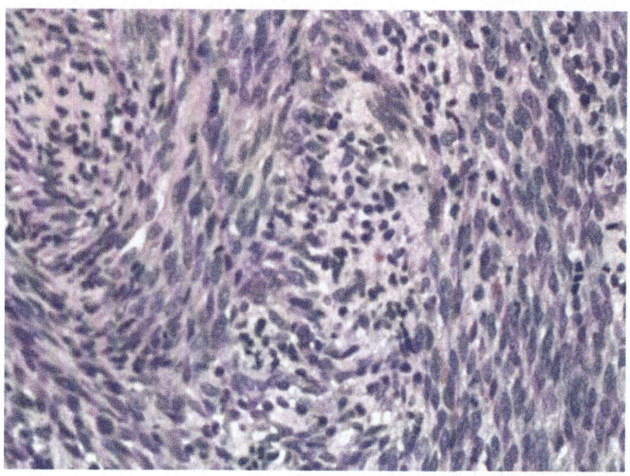

Fig. 2.45 Fibrosarcoma from humerus of a dog (Courtesy of Dr. Roy R. Pool, DVM, PhD, Emeritus Professor of Pathology UCVD and Texas A&M—USA)

Recurrences are not very frequent in case of fibrosarcoma of bone, while fibrosarcoma of soft tissues that may recur most frequent. Fibrosarcomas metastasizes most common to the lungs.

Macroscopic features of fibrosarcoma arise in different appearance with a swelling firm mass with different shapes and size, with circumscribed margins, with tan-gray-whitish areas, frequently collagenous and associated with necrosis and hemorrhages.

Cytologic features of smears or aspirates from fibrosarcoma of bone appear variable depending on the grade of the tumor. Low-grade fibrosarcomas are characterized by low cellular smears characterized by densely packed tissue fragments of tumoral spindle cells referred to fibroblasts in a background composed also by single cells with nacked nuclei. The general appearance is referred to a tumor composed by spindle cells not well differentiated and without a specific architecture. The grade of atypia of the spindle cells in low-grade fibrosarcomas may be variable with moderate nuclear atypia with granular chromatin and nucleoli not prominent. Usually mitotic figures are not present in smears of low-grade fibrosarcomas of bone while are detectable in the form from intermediate grade to high grade. In these forms also the oval nuclei may show different heterogeneous aspects of shape and volume and irregularities to nuclear membranes. In case of high-grade fibrosarcomas smears show dense higher cellularity, non-cohesive spindle cells with naked nuclei. General key cytological features of conventional fibrosarcoma of bone are resumed as:

- Different cellularity depending on the grade of malignancy of fibrosarcoma.
- Malignant cells arranged tissue fragments with fascicles or rare storiform pattern of pleomorphic spindle-shaped cells.
- Typical absence of multinucleated osteoclast-like giant cells.
- Necrosis and hemorrhages.
- Background of individual cells with naked nuclei.
- Anisocytosis and anisokaryosis variable and depending on grade of malignancy.
- Frequent presence of finely granular chromatin.
- Mitotic rate depending on grade of malignancy.

Image features of fibrosarcoma of bone are representative of a pure aggressive malignant osteolytic-destructive tumor with not well-defined limits and marginal sclerosis, interruption of the cortex, and absent-scarce periosteal reaction. The typical pattern reported from many authors may be also the moth-eaten or permeative with small fragments of residual bone. Pathologic fractures are frequent due to lysis of bone.

Histopathology is similar-identical to with that of fibrosarcomas of soft tissues with high cellular fiber-rich connective tissue with typical predominantly uniformly population of spindle-shaped tumor cell arranged in herringbone or fascicular pattern and mixed with variable amounts of collagen. Similar features can be identified also in other typology of bone tumors but fibrosarcoma always lack malignant osteoid and chondroid matrix as like in fibroblastic osteosarcoma or chondrosarcoma. The malignant cells in fibrosarcoma contain scarce eosinophilic pale cytoplasm, tends to form fascicles of different sizes (herringbone pattern) separated by different amounts of collagen fibers. Neoplastic cells arranged in bundles tend to intersect at right angles. Cellular differentiation of tumor cells and the related production of collagen tend to decrease consequently to the increase in the degree of malignancy. Usually is common considered that strong pleomorphism is not a characteristic of fibrosarcoma of bone.

Primitive fibrosarcomas of the oral cavity and of the head may have further diagnostic difficulties linked above all to the interpretation of the histopathological

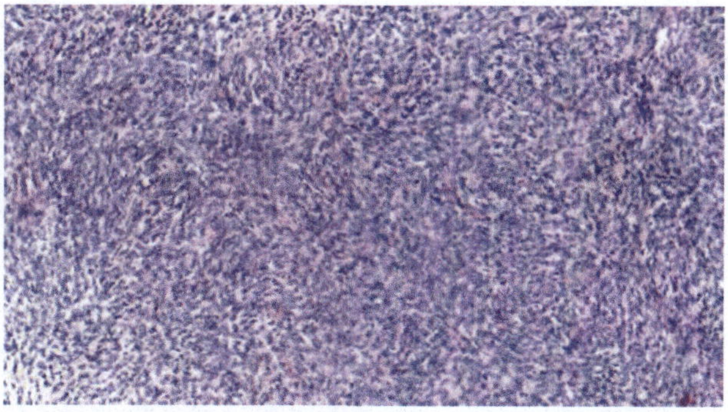

Fig. 2.46 Fibrosarcoma of bone

characteristics of the samples which, at times, does not allow us to understand whether the tumor originates from the bone or neighboring bone tissues and their exact classification may be related to behavior and infiltrative capacity of the tumor.

Regardless of the primary bone location (appendicular skeleton or axial skeleton), oncologists recommend subsequent treatments with therapeutic protocols mainly of radiotherapy after the removal/resection of the neoplastic tissue, which can contribute to raising the prognostic value of neoplastic bone disease.

Immunohistochemistry shows strong positivity to vimentin, weak to SMA (only in presence of myofibroblasts) and lacks immunoreactivity for epithelial and endothelial markers.

There are not data in literature for animals about genetics and little ones for humans with fibrosarcomatous tumors where have been detected several genetic imbalances where results do not permit to distinguish it from other types of bone sarcomas.

Ultrastructural studies demonstrate fibroblastic differentiation of malignant cells with abundant rough endoplasmic reticulum. Some of these cells have myofibroblastic differentiation and contain also cytoplasmic lysosomes characterizing fibrohistiocytes.

Differential diagnosis includes predominantly fibroblastic osteosarcoma that however contains always osseous matrix unlike in fibrosarcoma (Figs. 2.46 and 2.47).

There are also other forms of fibrosarcoma of bone, classified on the basis also of different primary and particular localization. These forms are represented by periosteal fibrosarcoma and maxillary or mandibular fibrosarcoma typical tumor of canine species. Periosteal fibrosarcoma of bone is usually a high-grade tumor arising superficially to periosteum and involving the surface of bone and frequently hardly differentiable from certain fibrosarcomas of soft tissues. Periosteal fibrosarcomas have been described in several species of animals, and most of these were localized in the head bones in generally well-differentiated forms with abundant collagenous stroma which can become more compact or also mixomatous. These tumors are characterized by frequent recurrences and rare metastasization processes which in the rarer cases may be detected first in lungs or lymph nodes.

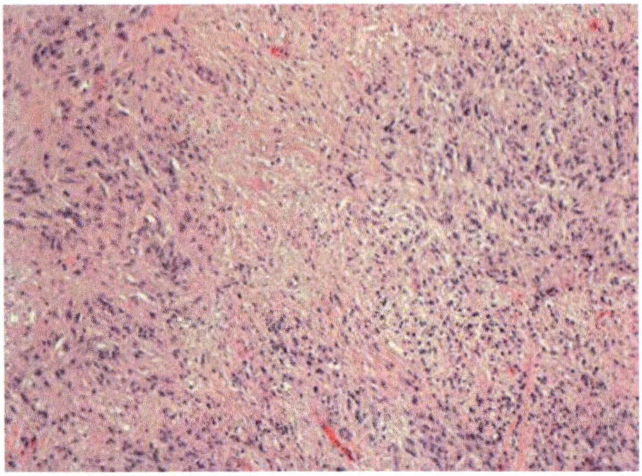

Fig. 2.47 Equine fibroblastic OSA

Maxillary and mandibular canine fibrosarcoma is a maxillary or mandibular low-grade deforming mesenchymal tumor of the face of middle-age dogs (average age of 7–9 years). These fibrosarcomas of bone are tumors with a slow growth that can cause a disfigure of the face, frequently also associated with loosening of the teeth and with large areas of osteolysis. Fibrosarcoma of maxilla/mandibular arising as a firm/soft mass firmly attached to the involved bone. Usually these tumors are less cellular of the other ones from soft tissues, but they are characterized by invasion of the below bone structures. Maxillary and mandibular fibrosarcomas are usually not capsulated tumors composed by low cellularity of well-differentiated fibroblasts, frequently also associated with mixed inflammation and edema. The index of metastasis is generally low especially when the grade of infiltration of surrounding tissues is not broad.

Fibrosarcoma of Bone

Definition: Fibrosarcoma is a high-grade pure fibroblastic and myofibroblastic (spindle cell) non-osteogenic sarcoma of bones producing collagen.

Etiology: predominantly unknown, but in some cases is possible a malignant transformation from pre-existing conditions. Treatment with radiation can represent a pre-existing condition.

Clinical presentations: enlarging mass with pain and frequent pathologic fracture.

Image Findings: Lytic (moth-eaten pattern) poorly circumscribed and aggressive lesion.

Macroscopic features: swelling firm mass with different shapes and size, with circumscribed margins, with tan-gray-whitish areas, frequently collagenous and associated with necrosis and hemorrhages.

(continued)

Microscopic pathology: is similar-identical to with that of fibrosarcomas of soft tissues with high cellular fiber-rich connective tissue with typical predominantly uniformly population of spindle-shaped tumor cell arranged in herringbone or fascicular pattern and mixed with variable amounts of collagen.

Differential diagnosis: Fibroblastic osteosarcoma.

2.12.6 Hemangiosarcoma

Hemangiosarcoma also called angiosarcoma or malignant hemangioendothelioma is a rare malignant bone tumor with angioblastic differentiation, originating from vascular endothelium of the bone and represents about 5% of all other canine bone tumors. Hemangiosarcomas in general have a wide range of morphological appearance and clinical aspects. Although hemangiosarcoma of soft tissues is relatively frequent in animals, the primitive hemangiosarcoma of bone represents a rare tumor in domestic animals and humans with significant diagnostic difficulties in differential diagnosis with other tumors characterized by the presence of blood, such as telangiectatic osteosarcoma. Of all animal species the dog is the most affected by skeletal hemangiosarcoma and it is also reported in cats, cattle, and horses. The breeds of dogs at greatest risk are represented by German Shepherds, Golden Retrievers, Labrador Retrievers, Boxers, and Great Danes.

No significant epidemiological data are available for hemangiosarcoma in animals as like as for etiopathological causes. In humans it seems that a small percentage of angiosarcoma of bones are associated with post-radiation treatment or bone infarction or skeletal hemangioma.

Hemangiosarcoma of bone is an aggressive tumor clinically characterized by a painful single or rarely multiple palpable and nodular mass. The mass generally is soft and very vascularized and hemorrhagic. Hemangiosarcoma of bone may cause bone destruction and also pathologic fracture.

The most frequent described skeletal locations in the dog are represented by long bones as femur, tibia, radius, and humerus, and less in other bones like vertebrae, scapula, pelvis, ribs, scapula, and maxilla. In literature there is also a rare description of hemangiosarcoma with involvement of sternum.

Imaging findings show peculiar aspects related to the differential investigative typologies represented by radiography, magnetic resonance, computed tomography, and bone scan. Radiographic findings usually give a wide range of info characterized by a destructive lytic mass, invasion of underlying bone with potential erosion of the cortex, and periosteal reaction and involvement of surrounded soft tissues. This tumor does not produce any type of matrix as appear in other primary bone sarcomas. Hemangiosarcoma of bone is also a metastatic tumor and when the metastasis involving other bones in polyostotic form it is impossible to detect the primitive localization of the tumor. The lungs represent the most common localization of the metastasis. Magnetic resonance delineates the distribution and extension of lesion with high-resolution details and computerized tomography can confirm the invasiveness behavior of the lesion. Bone scintigraphy, especially in cases where hemangiosarcoma occurs in an aggressive form, does not allow highlighting absorption of radionuclides if not sporadically along the margins of the cavitary blood lesions.

Macroscopy hemangiosarcoma looks like a nodular friable mass, hemorrhagic dark red, with scarce defined and irregular margins, usually involving surrounded soft tissues and also frequently associated with areas of necrosis. The cut surface shows spongy dark features, with abundant amounts of dark blood and blood clots. Fibrous tissue is always present and gives a firmer consistency to the mass tending to rubbery. Hemangiosarcoma of bone tends to invade and erode also the cortex and invading then soft tissues around. In rare cases a solid form of hemangiosarcoma of bone may be not have evidence of high vascularity.

Cytologic features of smears from hemangiosarcoma of bone have different features related to the grade of the tumor that can modify the intensity of cellularity always in a background rich in red blood cells. Hemangiosarcoma of low grade may containing a prevalence of red blood cells and low amounts of neoplastic cells while high-grade forms where is highest the cellularity with spindle, round, or polygonal cells, single or in clusters, with marked signs of anisocytosis and anisokaryosis. High grade of anisokaryosis is one of the distinctive cytological elements for differentiating cytologically hemangiosarcomas from hemangiomas. Nuclei are usually central, in pale cytoplasm with oval shape and with finely granular chromatin especially in the low-grade ones. Binucleated cells may be detectable in the smears from hemangiosarcoma. Granules of hemosiderin if present are not associated with neoplastic cells but with macrophages and often tumor cells can show signs of erythrophagocytosis. In the high-grade hemangiosarcoma of bone many nuclei are variable in sizes and have also smooth nuclear membranes with frequent invaginations.

Histologic features of hemangiosarcoma are represented by similar pattern of the same tumor of the soft tissues with vascular hematic cavities different in shape and sizes and always lined by atypical endothelial cells. The constant presence of malignant endothelial cells represents important features to differentiate hemangiosarcoma of bone from telangiectatic osteosarcoma where normally there are not endothelial cells. The neoplastic typical hematic spaces may have also the aspect and characteristics of big cavernous structures. I would like to remember that hemangiosarcomas from soft tissues can also be classified morphologically on the bases of the dimensions and distributions of hematic cavities in capillary and cavernous hemangiosarcoma. The abundant presence of blood can lead to the leakage from red blood cells of blood proteins that could accumulate in the neoplastic stroma and between the irregular septa of the fibrous tissue that tends to support the newly formed vessels, inducing a potential diagnostic failure and confuse this material with bone matrix. Frequently hemangiosarcomatous cells are characterized by malignancy and by an appearance similar to epithelioid cells that results predominant respect with the less frequent spindle cells. The hemangiosarcomatous cells have abundant deeply staining eosinophilic cytoplasm, sometimes vacuolated with vesicular nuclei with condensed chromatin, and one or more prominent nucleoli. Mitotic index is usually grading as medium-high. Collateral histopathological features in case of hemangiosarcomas are represented by sporadic formation of reactive bone, presence of extravasated erythrocytes, diffuse deposit of hemosiderin, and mixed inflammatory and intratumoral infiltrations. **Immunohistochemical phenotype** is characterized by positivity to several endothelial markers: Factor VIII (especially in well-differentiated cells), CD31, CD34, ERG (ETS-related gene (ERG), a transcription factor that has been linked to angiogenesis), and FLI1 (FLI-1 nuclear transcription factor, a nuclear marker of endothelial differentiation). In humans immunophenotype also includes positivity in individual tumor cells for SMA (Smooth Muscle Actin) and D2–40 (Lymphatic Endothelial Marker Monoclonal Antibody) and frequent positivity for keratin and EMA for malignant cells with epithelioid differentiation.

Most of hemangiosarcomas metastasize hematogenously and these forms are frequent letals with a surgery and limb amputation often unsuccessful due to its aggressivity.

Marconato e Del Piero suggested that once the diagnosis of hemangiosarcoma has been issued, it is always recommended to evaluate all the other bone structures, as well as the abdomen and chest to assess the presence of any metastases and to establish whether the bone lesion is also a metastatic or primary lesion, as the hemangiosarcomas occur more frequently in the spleen and liver and then metastasize also to the bones. Prognostic factors are generally poor even after amputation with average survival times of about 5 months.

Differential diagnosis refers mainly to forms of telangiectatic osteosarcoma and undifferentiated metastatic carcinoma.

Is suggested to consider hemangiosarcoma in differential diagnosis with those malignant epithelioid tumor of bone positive to keratin (Figs. 2.48 and 2.49).

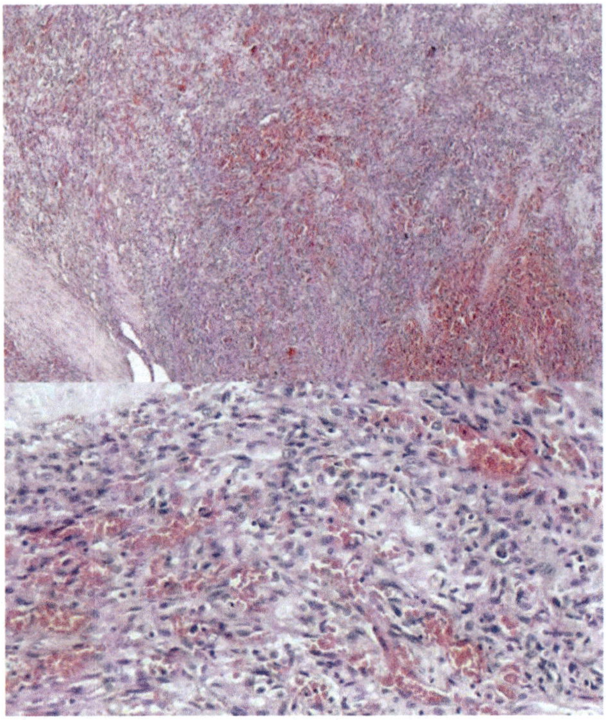

Fig. 2.48 Hemangiosarcoma, dog, mixed breed, 10 years old, F(S) (H&E, from author collection), 4x and 10x

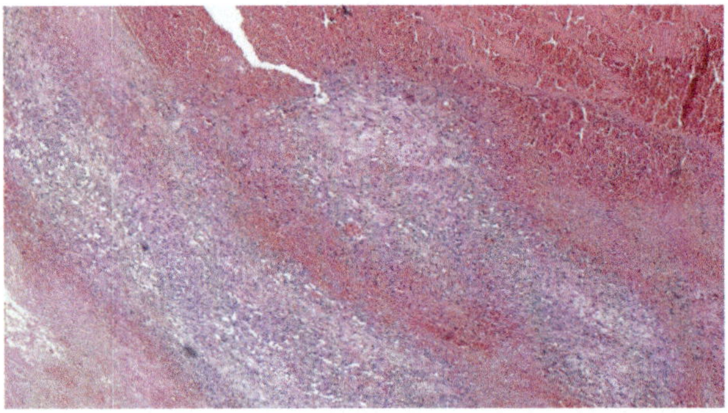

Fig. 2.49 Hemangiosarcoma, dog, mixed breed, 10 years old, F(S) (H&E, from author collection), 4x

Hemangiosarcoma of Bone
Definition: a rare malignant bone tumor with angioblastic differentiation and originating from vascular endothelium of the bone.

Etiology: predominantly described in humans with small percentage of angiosarcoma of bones are associated with post-radiation treatment or bone infarction or skeletal hemangioma or also contiguous to foreign materials.

Clinical presentations: characterized by a painful single or rarely multiple palpable and nodular masses lesion. The mass generally is soft and very vascularized and hemorrhagic. Hemangiosarcoma of bone may cause bone destruction and also pathologic fracture.

Image Findings: a wide range of findings with a defined destructive lytic mass, invasion of underlying bone with potential erosion of the cortex and periosteal reaction with involvement of surrounded soft tissues.

Macroscopic features: nodular friable mass, hemorrhagic dark red, with scarce defined and irregular margins, usually involving surrounded soft tissues and also frequently associated with areas of necrosis.

Microscopic pathology: vascular hematic cavities different in shape and sizes and always lined by atypical endothelial cells. The constant presence of malignant endothelial cells represents an important feature to differentiate hemangiosarcoma of bone from telangiectatic osteosarcoma where endothelial cells are not expressed.

Differential diagnosis: Telangiectatic osteosarcoma and undifferentiated metastatic carcinoma.

2.12.7 Giant Cell Tumor of Bone

Giant cell tumor of bone (GCT_b), also called osteoclastoma, is an intramedullary neoplasm with unpredictable behavior and controversial histogenesis. It has been considered as a benign neoplasm lesion but locally aggressive bone tumor. Giant cell tumor of bone consists histologically of three types of cells: mononuclear histiocytic cells, multinucleated osteoclast-like giant cells, and mononuclear neoplastic stromal cells that representing the main proliferating cell population. Large numbers of osteoclastic giant cells can be present in various tumors and non-tumors bone lesions and that these must not be confused with GCT_B. These include, among others: osteosarcomas, chondroblastomas, osteoblastomas, aneurismal bone cysts, and bone granulomas.

Mononuclear stromal cells are tumoral cells with phenotype similar to osteoblast stimulating formation of multinucleated osteoclast-like giant cells. Our previous comparative studies in cats and dogs and humans demonstrate different patterns of immunohistochemical expression in GCT_b with expression by mononuclear stromal cells of receptor activator NK-kB or RANKL ligand and expression by multinucleated osteoclast-like giant cells of receptor RANK that binds RANKL present on the surface of mononuclear stromal cells. It is defined an intramedullary well vascularized bone tumor with high potential of local malignancy and local recurrence. GCT_B derived from bone marrow stromal cells, and that arises mostly from the epiphyseal area of long bones and is associated frequently with pathologic fracture. Very few cases of GCT_B were described in domestic animals and only one with extraskeletal metastases was reported in a cat by Ferreras M.C. in 2005. Frequently GCT_B arises in the epiphysis and spreads out into metaphysis. Our experience with GCT_B from cats and dogs confirms this predilection for tibia and femur long bones. Very few studies and case reports are available in veterinary literature about GCT_B in domestic animals not sufficient to issue meaningful epidemiological data. In humans GCT_B represents the 5% of primary bone tumors and about 20% of benign bone tumors with a frequency of approximately 1/million/year. Campanacci described also in humans a sarcomatous transformation change in less than 5% of all giant cell tumors and in contrast with most others tumor of bone a little preference to females than males. In our experience we found that adult mature cats are more affected than dogs.

Etiology recognizes that multinucleated osteoclast-like giant cells in GCT_B are not neoplastic and that the mononuclear cells represent the tumor cellular components that may show different grades of atypia and also mitotic activity. In humans secondary malignancies in GCT_B follow radiation therapy in several cases but in animals there are no data reported related to etiology. Remain interesting the considerations about different levels of activation of RANK/RANKL/OPG system in all cellular components described in these tumors.

Clinical issues report a prevalence of primitive localization in epiphyseal-metaphyseal areas of long tubular bones rarely involved the diaphysis. In humans have been described uncommon sites in vertebral bodies, craniofacial bones, patella, hands, and feet. GCT_B usually is solitary is very rarely multifocal. WHO classification of tumours of soft tissue and bone recognizes in humans two different ICD-O Code (International Classification of Diseases for Oncology), one for giant cell tumor of bone (9250/1) and one for malignancy in giant cell tumor (9250/3) attributing an official identity even to the rare malignant forms of this tumor. Clinical presentation of GCT_B is usually characterized by pain, swelling, and less frequent with pathologic fracture and hemarthrosis. Moderate pain, swelling, limitation of joint movements, and increase of local temperature could represent the first symptom that usually could be referred to joint due to the primary locations of the tumor.

Imaging in radiographic findings shows an expansile, large, lytic, eccentric, lobulated, intramedullary, osteolytic lesion with small limit of transition lined by sclerotic bone. The typical radiologic appearance also called a "soap bubble or puddle on the sand" is made by expanded and thin cortical bone. Sometimes tumor extend to soft tissues and results covered by proliferation of thin sclerotic tissue. Usually no matrix calcification is identified within the tumor. GCT_B may expand bone and cause reactive periosteal bone formation, with thin reactive discontinuous shell of new forming bone. In medullary cavity usually may be involved with well-defined margins and "moth-eaten" appearance. In humans a data of particular interest is represented by the effects caused by the therapy using

RANKL inhibitors, which, after treatment, determine an increase in sclerosis and bone formation phenomena, which are easily detectable on radiographic examination. CT scan reveals that GCT_B is generally lytic, with well-defined borders, homogeneous, and that can also contain cystic cavities. Also, CT shows a thin shell of reactive bone not dense and not compact sometimes with several interruptions characterized by radiolucent perforations. MR findings can give also other details related to T1 or T2 protocols of application. T1 results better to evaluate the tumor in its intramedullary components, while T2 is preferred for the evaluation of extraosseous features of tumors with large size and to identify and evaluate also cystic cavities containing various amounts of fluids. Also isotope bone scan may be useful to establish different ranges of bone uptakes of technetium and angiography that can confirm the typical pattern of hypervascularity of GCT_B mixed with other stromal avascular cystic and necrotic components.

Gross pathology of GCT_B shows the general features of a tumor well defined and demarcated by a capsule or reactive bone or fibrous tissue. The mass is hemorrhagic, friable, and homogeneous yellow jelly (xanthomatous changes), red-brown in color, with a range from 4–5 to 10–15 cm in dimension. The tumor commonly is also necrotic (pale yellow in color) and cystic, with whitish areas of focal fibrosis and hypervascular and rarely bleeding. When tumor extend through the cortex show pushing margins. Instead when rarely expand in subchondral bone and reach the deep surface of articulation it can penetrate also in articular surface. When GCT_B acquires marked malignant characters the gross appearance is similar to any other high-grade sarcoma generally involved also the neighboring soft tissues.

Cytologic features are represented by moderate to abundant cellularity mixed with tissue fragments of different origins in abundant blood background. Low magnification view reveals a typical cellular population characterized by single or cluster of mononuclear spindle-ovoid cells mixed with multinucleated osteoclast-like giant cells. Often the clusters of mononuclear neoplastic cells appear surrounded by the other multinucleated osteoclast-like cells in a characteristic pattern of giant cell tumor of bone called "checkerboard." The mononuclear stromal cells have a typical spindle-oval shape with moderate pale eosinophilic cytoplasm and mild eccentric nuclei with round or ovoid shape.

Unlike of normal osteoclasts the multinucleated osteoclast-like giant cells in GCT_B contain a much larger number of nuclei represented by more than 30 up to almost 100. These nuclei are regular and similar in size, shape, and chromatin appearance. The presence of aneurysmal bone cyst, fibrosis, hemorrhages, necrosis, and xanthomatous degeneration can represent the main cause to obtaining not representative smears of the tumor. Some problems can be found with older cases where there are bigger amounts of fibrosis which can result in smears very low in cellularity. Anyway the cytological key features of giant cell tumor of bone are resumed in:

- Variable amount of cellularity in a bloody background;
- Two different main cell population represented by mononuclear neoplastic stromal cells (single or in syncytial cytoplasm clusters) and osteoclast-like giant cells;

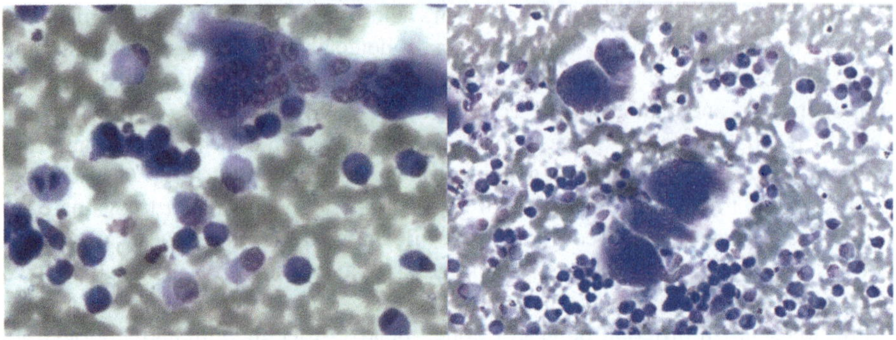

Fig. 2.50 Cytology from a case of GCT_B in a cat where it is possible to detect easy the presence of mononuclear stromal cells and multinucleated giant cells, typical of GCT_B (from author collection)

- Checkerboard pattern is represented by tissue fragments surrounded by a "layer" of adherent multinucleated osteoclast-like giant cells;
- Mononuclear cells show short spindle-oval shape;
- Similar appearance of size and shape of nuclei of mononuclear stromal cells and multinucleated; osteoclast-like giant cells;
- Nuclei show condensed chromatin;
- Multinucleated osteoclast-like giant cells have frequently more than 30–40 nuclei;
- Mitotic figures present almost exclusively in mononuclear stromal cells (Fig. 2.50).

Histopathology of GCT_B is characterized by different histotypes mixed with to fibrotic, necrotic, hemorrhagic, and xanthomatous lesions and it is advisable to focus histopathological observation on the areas in which the cellular characteristics of the tumor are best preserved to avoid diagnostic risks. Histologic features in GCT_B referring to a cellular tumor moderately-well vascularized and composed of large number of multinucleated osteoclast-like giant cells mixed with numerous round-oval-spindle mononucleated stromal cells also mixed with a smaller number of mononuclear histiocytic cells. Stromal mononuclear cells are the main proliferating tumoral cells population. Multinucleated osteoclast-like giant cells contain a variable large number of nuclei generally always more than 20–30 but also up to 100 that appears identical to those of the stromal mononuclear cells. Giant cells have plump eosinophilic cytoplasm, vesicular round to oval nuclei with prominent nucleoli. Mononuclear stromal cells are diagnostic with specific grow of its in syncytium and indistinct borders, round-oval to spindle shaped, pale moderate eosinophilic cytoplasm and round-oval nuclei with prominent central nucleoli. Mononuclear stromal cells are the most active cellular clones and usually show different degrees of atypia associated with a presence of frequent mitotic figures. In some cases of GCT_B is described a development of aneurysmal bone cyst with a histological appearance of sponge-like feature and with cystic spaces rich in blood and separated by connective septa where it is possible to detect the characteristic cellular

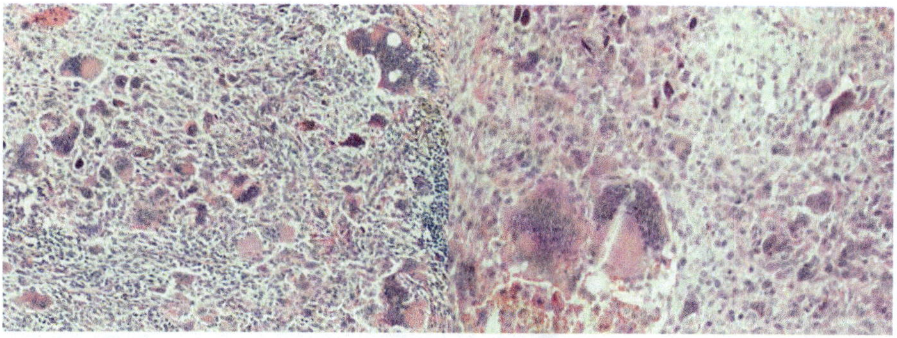

Fig. 2.51 H&E stain of a case of GCT_B of a cat. Note, the abnormal presence of multinucleated giant cells characterizing this neoplastic form at several magnifications (from author collection)

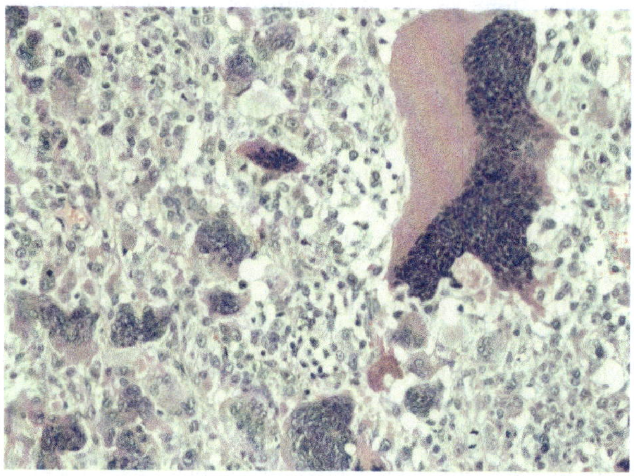

Fig. 2.52 H&E suggestive histological image of a case of feline GCTB with also one big multinucleated giant cell containing a very large number of nuclei (from author collection)

population of the tumor and sporadically also foci of osteoid and chondroid matrix more detectable at the periphery of the tumor but also in the central areas. Reactive woven bone formation may be also frequent. In some cases of GCT_B it is possible to detect diffuse areas of fibrosis, scarce amount of collagen and reticulin, and histiocytic cells in scattered aggregates with foamy histiocytes with pyknotic nuclei. Necrosis may be frequent and extensive as like as hemorrhages and consequently fibrous tissue, hemosiderin, foam cells, and diffuse inflammatory infiltration of different grades. In human medicine primary malignancy in GCT_B was described in areas with large amounts of very pleomorphic mononuclear cells (Figs. 2.51, 2.52, and 2.53).

Immunohistochemistry demonstrates of different markers with multinucleated osteoclast-like giant cells with a similar immunoprofile to macrophages and stromal mononuclear tumoral cells with an osteoblastic phenotype staining for RANKL and nuclear staining also for p63. Multinucleated osteoclast-like giant cells express receptor for vitronectin (CD51), also called integrin alpha chain V or vitronectin

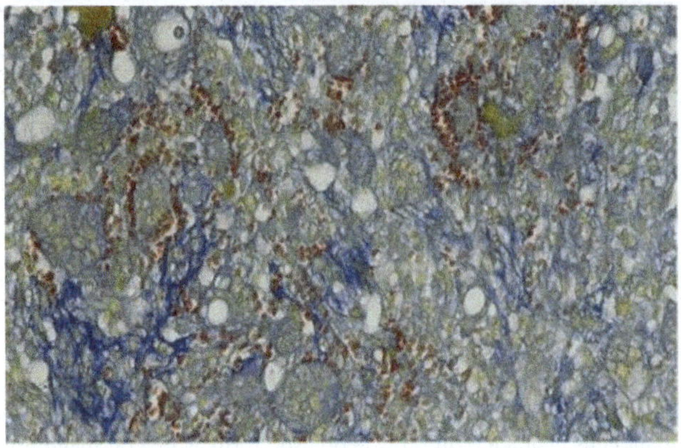

Fig. 2.53 Mallory's trichrome of GCT_B from a cat (from author collection)

receptor-alpha chain, a heterodimeric integral membrane protein with an extracellular domain, a transmembrane region, and cytoplasmic domain. Multinucleated osteoclast-like giant cells express receptor have also a limited range of macrophagic markers including CD45 (transmembrane protein tyrosine phosphatase located on most hematopoietic cells), CD33 (myeloid cells differentiation marker), CD68 (routinely marker of inflammation for monocytes/macrophages). Multinucleated osteoclast-like giant cells express receptor do not react with CD14, CD163, or HLA-DR. These cells show a strong positivity also for cathepsin K and tartrate-resistant acid phosphatase.

Differential diagnosis enclosed several other forms of bone tumors or bone lesions. The main of these are represented by malignant fibrous histiocytoma where the multinucleated osteoclast-like giant cells are less and scattered and mononuclear stromal cells have spindle shape with cells distributed mainly at the periphery of the tumor. Other differential diagnoses are with chondroblastoma, giant cell osteosarcoma, and aneurysmal bone cysts. Chondroblastoma is characterized by mononuclear cells with features of the nuclei not resembling the nuclei of multinucleated osteoclast-like giant cells as instead happen in GCT_B and usually in chondroblastoma mononuclear stromal cells stains for S100 and are negative for p63. Giant cell osteosarcoma in very infiltrative tumor contains malignant mononuclear osteoblastic cells and major amounts of bone matrix. Aneurysmal bone cyst is much more difficult to differentiate from GCTB especially if the latter manifests itself in the cystic form where however the cystic areas may have characteristic morphologic features.

Surgical therapy for GCT_B generally provides for amputation in case of appendicular locations, while surgical resection followed by radiotherapy is more recommended in cases of axial tumor locations (Figs. 2.54 and 2.55).

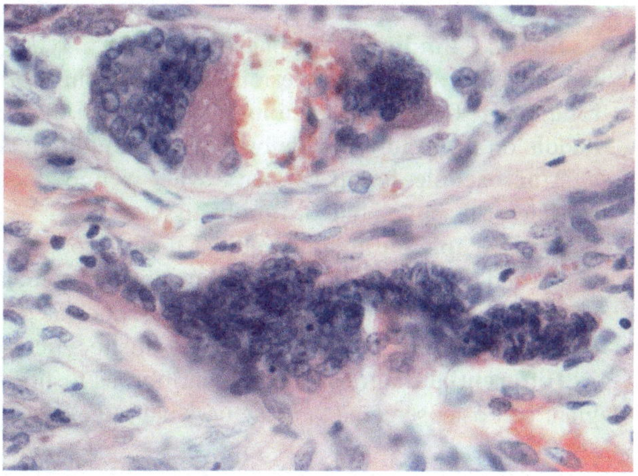

Fig. 2.54 H&E stain of a case of GCT_B of a cat. Note, the abnormal presence of multinucleated giant cells characterizing this neoplastic form at several magnifications (from author collection)

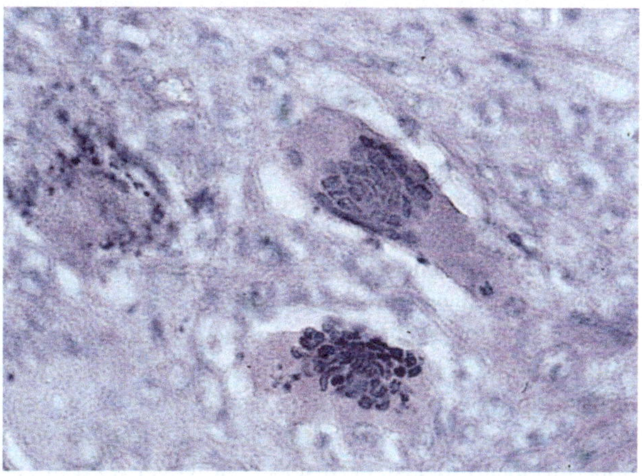

Fig. 2.55 H&E stain of a case of GCT_B of a cat. Clear presence of multinucleated giant cells characterizing this neoplastic form at several magnifications (from author collection)

Giant Cell Tumor of Bone

Definition: an intramedullary neoplasm with unpredictable behavior and controversial histogenesis. It has been considered as a benign neoplasm lesion but locally aggressive bone tumor.

Etiology: in human secondary malignancies in GCT_B follow radiation therapy in several cases but in animals there are no data reported related to etiology.

Clinical presentations: usually characterized by pain, swelling and less frequent with pathologic fracture and hemarthrosis.

(continued)

Image Findings: expansile, large, lytic, eccentric, lobulated, intramedullary, osteolytic lesion with small limit of transition lined by sclerotic bone. The typical radiologic appearance also called a "soap bubble or puddle on the sand" is made by expanded and thin cortical bone.

Macroscopic features: mass with well defined and demarcated by a capsule or reactive bone or fibrous tissue.

Microscopic pathology: cellular tumor moderately-well vascularized and composed of large number of multinucleated osteoclast-like giant cells mixed with numerous round-oval-spindle mononucleated stromal cells also mixed with a smaller number of mononuclear histiocytic cells.

Differential diagnosis: malignant fibrous histiocytoma, chondroblastoma, giant cell osteosarcoma, and aneurysmal bone cysts.

2.12.8 Multilobular Tumor of Bone

This is likely the most common bone tumor occurring in the head of dogs. Its initial name, calcifying aponeurotic fibroma, was influenced by a consulting physician pathologist who thought the nodules were in the fascial attachment of nuchal musculature to the caudal border of the skull of a dog. To him the nodules in the cranial fascial attachment resembled nodules in some cases of plantar fasciitis in man. A second nodular skull lesion was shown to another physician pathologist who called the lesion, chondroma rodens, because it acted like a rodent ulcer as it ate its way into the adjacent bone surface. Pool after reviewing 25 examples of this tumor in the files of UC Davis, SVM observed three morphologic groups based upon the major nodular pattern in each specimen. He initially recognized multilobular osteoma, multilobular chondroma, and multilobular osteochondroma, all of which according to clinical histories in their files were locally aggressive lesions and none produced metastatic disease. However, two of several new clinical cases that he examined demonstrated malignant behavior. Both calvarial lesions metastasized to the lungs and the metastases recapitulated the nodularity of the primary tumors. After this experience he applied an inclusive term, multilobular tumor of bone (MLT) to this category of bone tumors, since at time of initial biopsy, the potential for malignant behavior could not be ruled out. Radiographs that were available for review of many MLTs supported the interpretation that most of the MLTs appear to straddle or be associated with suture lines of the calvarium and even at the temporozygomatic suture. Subsequently, several nodular lesions with overlapping histologic features to MLTs of the calvarium were centered on the palatomaxillary suture line and a smaller number arose from the incisivomaxillary suture line. These findings support a concept that many MLTs arise from a disturbance of suture lines in the head of the dog. Other nodular lesions occurring on bone surfaces of the skeleton do not reflect the same precise nodular arrangement found in MLTs that have a thin outer lining resembling a fibrous periosteum that overlies as a slightly

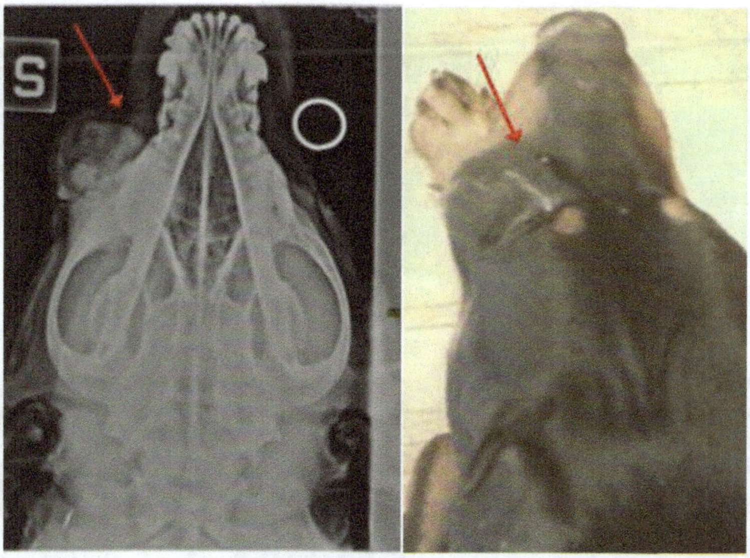

Fig. 2.56 Multilobular tumor of bone. Gross (left) and radiologic (right) appearance of the tumor. Radiography shows the dorsoventral projection of the skull and zygomatic area with the mass (arrows). (Leonardi et al. 2014a)

thicker deeper layer that can form cartilage, bone, or both in different nodules of the same mass. Many of those initially having cartilaginous centers may undergo endochondral ossification. Those tumors arising in the suture lines at either end of the hard palate form less precise nodules as compared to those of the skull. Malignant progression to chondrosarcoma and osteosarcoma may occur in some nodules often located at the borders of a MLT.

MLT of bone more often affects middle-aged dogs of larger breeds having a medium age of 8 years. The few lesions said to be examples of MLR in cats and horses are not well documented as compared to MLTs of dogs.

In our experience we described a case of the zygomatic bone in a 10-year-old mixed and medium-sized breed dog (Fig. 2.56).

Imaging features of multilobular tumor of bone revealed a single nodular lesion with unclear smooth borders localized to the bone structures and characterized by soft tissues with diffuse areas of multifocal calcifications. One other feature may characterize a "popcorn ball" appearance related to different grades of granular aggregates frequently associated with localized mild geographic lytic lesions of adjacent bones that become large when tumor acquires more malignant characters (Figs. 2.57 and 2.62).

CT axial image and sagittal reconstruction evidence the complete drilling of the calvarium with a primitive invasion of the cranial cavity.

The volume rendering of the CT shows the tridimensional expansion of the mass out of the skull.

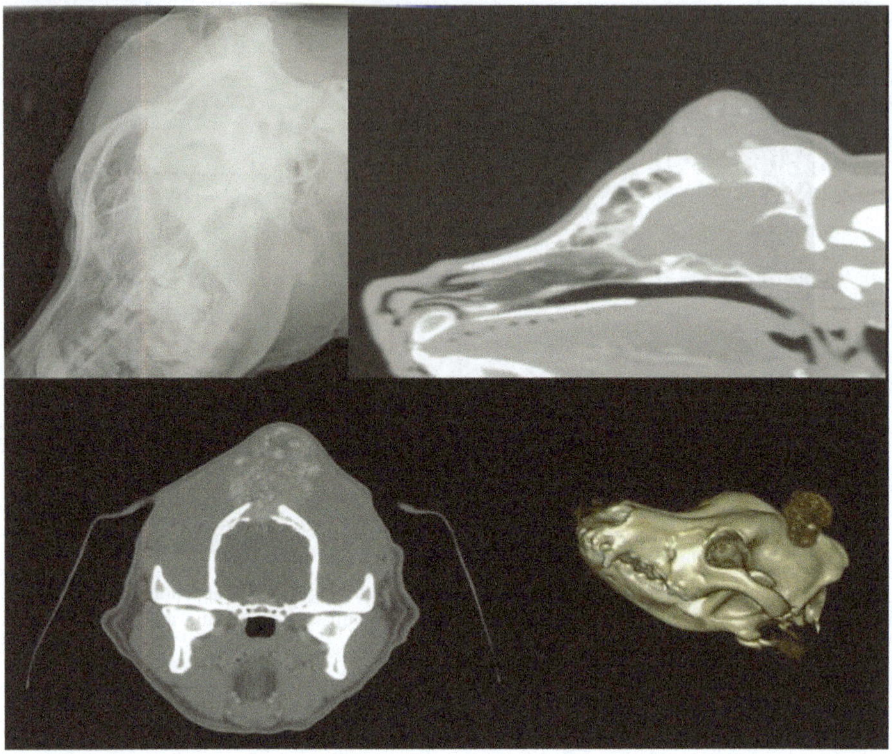

Fig. 2.57 Multilobular tumor of bone in a dog, mixed breed, female, 5 years old. Lateral X-ray view shows a large bone defect in the upper part of the skull, caudally to the frontal sinuses. In the soft tissues is visible a mass with a well-defined granular mineralized aspect

(Courtesy of Dr. Giovanni Angeli—Department of Veterinary Medicine, University of Perugia)

Clinically the tumor shows a small soft tissue painless mass with slow growing characters, firm but not with hard texture. Multilobular tumor of bone do not recur frequently after surgery in dogs when margins are widely resected. When the margins are incomplete the index of recurrence may increase. When the tumor becomes malignant and metastasizing, the lung represents the most frequent stricken organ. Dogs with recurrence could metastasize more easily. Anyway, the metastatic forms of multilobular tumor of bone have a slower growth with respect to osteosarcoma or chondrosarcoma also if these may arise also within multilobular tumor of bones.

Grossly multilobular tumor of bone is usually compact, gray-white to gritty infiltrating the soft tissues, tendons, muscle, fat, vessels and nerves, without showing well-defined borders. Associated with this condition is always present a diffuse pattern of fibromatosis that tends to form septa of different thickness.

Cytological features show low cellular smears with clusters of mesenchymal spindle to oval cells with a background of cartilaginous or bone matrix. Neoplastic cells have basophilic cytoplasm, central nuclei with prominent nucleoli and, occasionally, binucleate and multinucleated osteoclastic cells with usually low grade of mitotic index that can increase with the increase of the malignant grade of the tumor. Definitive diagnosis is not easy and the histological examination always remains the investigative method to reach the diagnosis of certainty.

Histopathology of multilobular tumor of bone is characterized by a feature of multiple lobules of neoplastic cells contoured by septa of different thickness and composed by large number of spindle-shaped mesenchymal cells. The main lobules contain different amounts of hyaline cartilage and cartilaginous or osseous matrix alone or in different ratio. Not infrequently is possible detecting polyhedral hyperchromatic shaped cells similar to osteoblasts localized in a kind of linear aggregates in the central area of matrix. In the most benign forms mitotic index is always low, while in multilobular tumor of bone with clear signs of malignant transformation where the atypical cellular mitosis increases in number. Foci of calcifications and chondroid differentiation are not uncommon inside the tumor stroma. Multinucleated giant cells resembling osteoclasts may be detected during histopathological investigations as like as hemorrhages, necrotic and peripherals inflammatory areas.

Concerning the prognostic factors of multilobular tumor of bone its regarding mostly the highest malignant form where the metastatic spread has the same biological features of other malignant mesenchymal tumors as osteosarcoma and chondrosarcoma but the few cases described above all in dogs have reported longer survival times for metastatic multilobular tumor of bone than the other more aggressive forms of osteosarcoma and chondrosarcoma (Fig. 2.58).

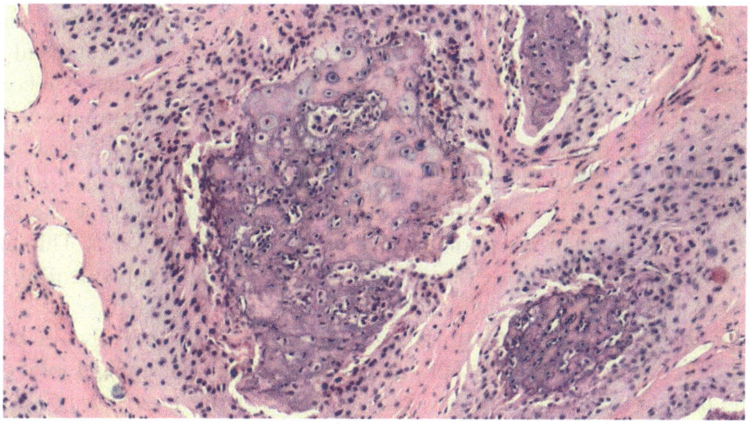

Fig. 2.58 Multilobular tumor of bone. Multilobular pattern of irregularly shaped and sized cartilaginous islands separated by fibroblast-like cells. H&E stain, 4x (from author collection)

Multilobular Tumor of Bone

Definition: a bone tumor with a multilobular pattern and containing islands of cartilage and bone at different degrees of differentiation and separated by connective tissue septa of spindle cells.

Etiology: unknown etiology and unknown histogenesis.

Clinical presentations: small soft tissue painless mass with slow growing characters.

Image Findings: a single nodular lesion with unclear smooth borders localized to bone structures and characterized by a soft density tissues with diffuse areas of calcifications multifocal (popcorn balls feature).

Macroscopic features: compact, gray-white to gritty infiltrating the soft tissues, tendons, muscle, fat, vessels, and nerves, without showing well-defined borders.

Microscopic pathology: multilobular tumor of bone is characterized by a feature of multiple lobules of neoplastic cells contoured by septa of different thicknesses and composed by large number of spindle-shaped mesenchymal cells.

Differential diagnosis: chondroma and chondrosarcoma, osteoma and chondroblastic osteosarcoma, fibroma, fibrosarcoma.

2.12.9 Synovial Sarcoma

Synovial sarcoma is a rare malignant mesenchymal tumor of joints having a wide spectrum of histological patterns and where the sarcomatous cells arising from synovioblastic mesenchyme maintain different degrees of epithelial differentiation as like as in normal synovium with usual stain positive for keratin. Synovial sarcoma has been described in different animal species and in dogs arises frequently in stifle, elbow, shoulder, carpus, and hip joints. Also, in other species as in cats it is classified also as synovial sarcoma but after very deep evaluation of the histological aspects and exclusion of other potential similar histiocytic tumors. In humans it accounts for 6–10% of all sarcomas of soft tissues, and through the joint tumors it follows malignant fibrous histiocytoma, liposarcoma, and rhabdomyosarcoma. In dogs synovial sarcoma shares the frequency with histiocytic sarcoma of joints. The diagnosis based the main evaluations on primitive localization inside the joint and on identification of main morphological aspects of malignant cells. This tumor is usually para-articular but also extra-articular with frequent involvement of tendons sheaths and adjacent extra-articular soft tissues.

Symptoms and clinical features refer a "deep" mass (not always palpable) centered on the joint associated with other clinical feature of pain and joint contracture. The initial growth often is slow but tends to increase over time. Tumor has a round to lobulated shape and frequently it results firmly attached to joint capsule or tendons.

Imaging features show frequently a tumor with foci of calcification-ossification. In synovial sarcoma different firm lobules usually tends to fill the joint cavity and the tumor is frequently associated to hemorrhages, necrosis, and mucinous dense fluid inside the joint cavity, with frequent involvement of adjacent bones and characterized by superficial erosions, reaction of periosteum, and focal areas of lysis.

MRI revealed a mass not homogeneous and in CT and MRI angiography with contrast the features of the tumor results in a well-defined hypervascular roundish or multilobulated mass with infiltrative margins.

Macroscopic features reveal an infiltrative usually mass, which frequently appears small due to the particular primitive localization, and due to also the slow growth. The tumor frequently is well circumscribed but can infiltrate often the adjacent bones and muscles. Cut surface is quite regular, also due to the cellularity of the tumor, and often associated with necrosis, hemorrhages, and myxoid changes, which can cause color variations of the cut surface with a range from greyish-to-greyish brown or yellowish. Synovial sarcoma frequently is multicystic with cysts filled of mucinous fluid.

Cytologic features: cytology of synovial sarcoma has a wide variety of patterns with high cellular smears composed by a mixture of single cells or clusters of cells with a mixture of straggler tissue fragments.

Neoplastic cells are frequently broken with only residual of nuclei, but generally they have round to ovoid shape, sometimes spindle especially in poorly differentiated synovial sarcomas. Nuclei have little amounts of chromatin and no clear nucleoli. One common cytological feature associated to this typical biphasic pattern is the sporadic presence of small glandular round to oval structures that can show also a lumen.

Histologically synovial sarcoma can have histologic features of several fibrous tumors. Synovial sarcoma contains a biphasic cellular pattern represented by synovioblastic and synoviocytes components with tendency to form epithelioid cells and spindle-shaped fibroblast like cells, making the sclerotic cellular component. The spindle-shaped cells tend to arrange in different types of aggregates for size, shape, and number of cells: cords, sheets, packed packages. Spindle cells appear usually as plump, uniform, with small or moderate cytoplasm and oval hyperchromatic nuclei. Often these cells may also resemble a pattern similar to hemangiopericytoma that represents one other potential differential diagnosis. The epithelial cells are characterized by cuboidal, ovoid cells forming different tubular or ductal structures, also papillary. Rarely epithelial cells can compose aggregates of epidermoid or squamous cells. Neoplastic cells show ovoid, angular, or fusiform shape with hyperchromatic cytoplasm, eccentric nuclei, and variable mitotic index related to the grade of malignancy of the cells. Frequently synovial sarcoma can be associated with multinucleated giant cells whose presence is not pathognomonic for this type of tumor. Mitotic index is related to the degree of malignancy of the tumor, i.e. generally higher in case of high-grade malignancy. Pool R.R. report also some synovial sarcoma diagnosed with a predominantly monophasic pattern, which necessarily involves a differential evaluation with poorly differentiated forms of fibrosarcoma or metastatic carcinoma.

Histochemistry and Immunohistochemistry: I consider appropriate to make an excursus on some peculiar histochemical characteristics useful in the diagnostic investigations for synovial sarcomas. The use of reticulin stain is useful to identify the biphasic pattern because the positive stain result always abundant corresponding to spindle cells aggregates and negative for epithelial cells. On the other hand, the epithelial cells aggregates/structures where is possible to detect secretions stains P.A.S. positive.

Immunohistochemical stains for synovial sarcomas permit to detect epithelial cells and rare spindle cells stain positive for cytokeratins and epithelial membrane antigen (EMA). Synovial sarcomas express EMA and several keratins as 7, 8, 14, 18, 19, and 20. There are cytokeratins expressed by epithelial cells but also by other spindle cells, like cytokeratins 7, 8, 18, and 19. Human WHO reports a positivity of these last markers in 70–80% of tumors. Only spindle cells stain for Vimentin and often also for S-100 protein, while CD34 (a transmembrane phosphoglycoprotein, often used to quantify the number of hematopoietic stem cells) stain rarely especially in monophasic type (Figs. 2.59 and 2.60).

Interesting ultrastructural feature of the epithelial cells in synovial sarcoma refers the presence on the surface of these cells of villous filopodia and several intercellular

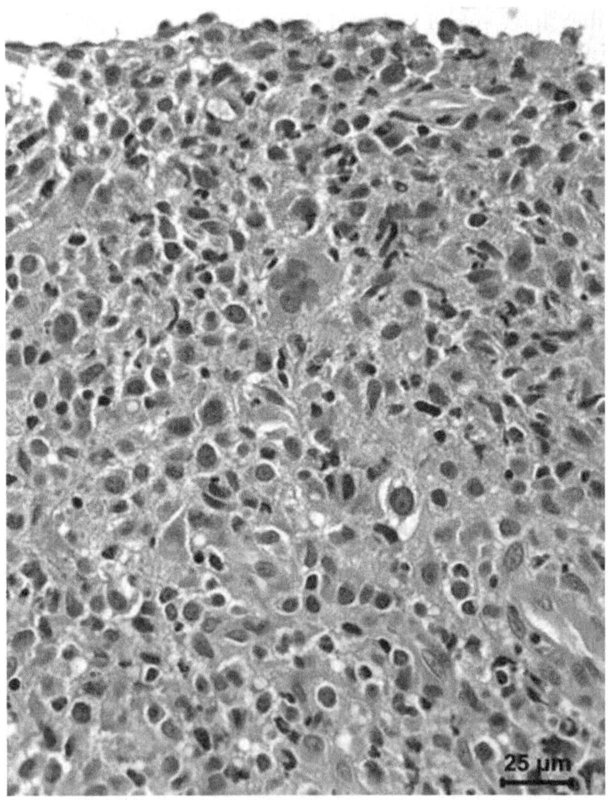

Fig. 2.59 Synovial sarcoma with lots of histiocytic appearing cells in the stifle joint of a dog (Courtesy of Dr. Roy R. Pool, DVM, PhD, Emeritus Professor of Pathology UCVD and Texas A&M—USA)

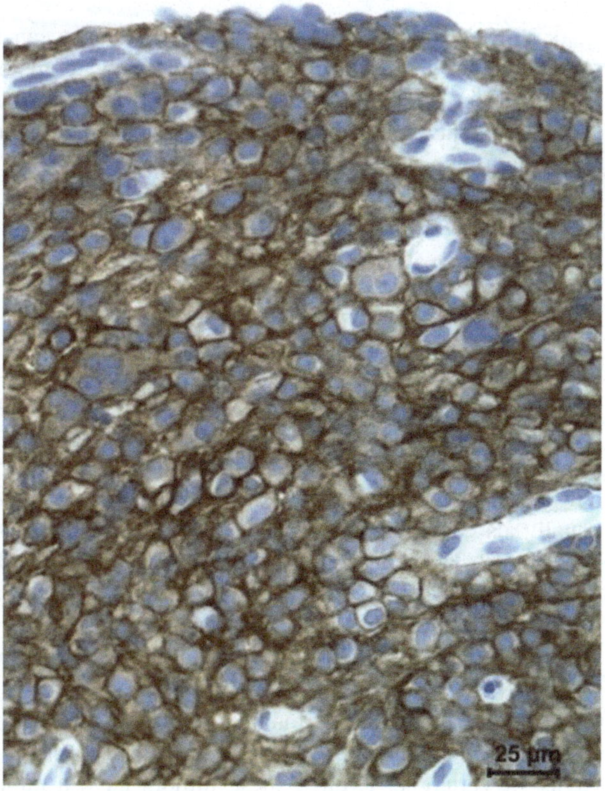

Fig. 2.60 CD18 immunohistochemical stain from the above slide demonstrating the histiocytic nature of most of these tumor cells in this particular synovial sarcoma (Courtesy of Dr. Roy R. Pool, DVM, PhD, Emeritus Professor of Pathology UCVD and Texas A&M—USA)

junctional structures. Campanacci describes other similar characteristics also for spindle cells that permit to consider that spindle cells may also undergo epithelial differentiation.

Synovial Sarcoma

Definition: a rare malignant mesenchymal tumor of joints having a wide spectrum of histological patterns.

Etiology: unknown in animals. In humans has specific chromosomal translocation t (X; 18) (p11; q11) that leads to formation of a SS18–SSX fusion gene.

Clinical presentations: a mass centered on the joint associated with other clinical feature of pain and joint contracture.

Image Findings: a mass with different firm lobules usually tends to fill the joint cavity.

Macroscopic features: infiltrative mass with cut surface quite regular and associated often with necrosis, hemorrhages, and myxoid changes.

(continued)

Microscopic pathology: synovial sarcoma can have histologic features similar to several fibrous tumors. Synovial sarcoma contains a biphasic histological pattern represented by synovioblastic and synoviocytes components with tendency to form epithelioid cells and spindle-shaped fibroblast like cells forming the sclerotic cellular component.

2.12.10 Liposarcoma

Liposarcoma of bone is a tumor of domestic animals with specific fat phenotype and that arises with adipose origin usually within the bone or in its surface. Adipocytic tumors of bone are very rare despite the abundant presence of fats in bone marrow. The benign counterpart of liposarcoma is lipoma that can arise in the bone intramedullary, intracortical, or parosteal (surface) and is characterized by mature adipocytes. Other few words about lipoma describing it as intraosseous lesion, usually also asymptomatic, and very quiescent tumor sometimes similar and confusing with atrophic or degenerative changes. In dogs and cats, soft tissue liposarcomas originate from the lipoblasts of the subcutaneous tissue and for these, foreign body pathogenesis has also been supposed.

Liposarcoma of bone is a tumor mainly described in dogs and cats and characterized by osteolytic infiltrative bone lesion, and usually is a high-grade bone tumor developing primarily in the long tubular bones with a particular predilection in dogs for femoral head and for lower frequency in humerus, tibia, radius, and marrow cavity of vertebrae. Even if these tumors have a rather aggressive behavior, the phenomena of metastasis represent events considered rare.

Usually, liposarcoma of bone arising as a painful large, soft-firm mass, with irregular and not well-defined margins, infiltrative and sometimes lytic and often myxoid. At CT and MRI investigations appear clear the main lipid components of the mass. The cut surface may reveal a lobulated tumor, yellow or yellow-tan in color.

Cytology features of well-differentiated liposarcoma are characterized by a predominant component of adipose tissue similar to what happens in case of lipomas. An interesting macroscopic aspect of the smear prepared with material represented by a suspect of adipocytic tumor of potential lipid origin, is characterized by a "greasy" smear. Cytological examination of a liposarcoma reveals the presence of a background not very rich in cells but in which roundish, sometimes polygonal cells with abundant cytoplasm and frequently vacuolized referable to adipocytes, are detectable. The vacuoles often appear clear and can vary in shape and size and can pushing nuclei to the poles of cells. Nuclei are hyperchromatic, without prominent nucleoli and also without atypical mitoses. In well-differentiated liposarcomas the presence of lipoblasts may be useful to support the final diagnoses.

Histologic features: histopathology of liposarcoma of bone is similar to those of soft tissues. Liposarcoma is composed of lipocytes usually aggregates in lobules separated by collagen tissue. The lipocytes are usually well differentiated as a signet ring multivacuolated cells with hyperchromatic nuclei. Some cases show a fibromatous similar pattern with cellular components characterized by numerous not well-differentiated spindle cells. In humans this type of liposarcoma is classified as sclerosing liposarcoma but this type is not recognized officially in veterinary medicine where the differences included well differentiated, poor differentiated, or myxoid types (this last form is officially recognized also in humans as well as pleomorphic type). Special stains do not seem particularly necessary since generally the classical methods of histopathological staining allow making morphological diagnosis relatively easy. If necessary, histochemical stains for lipids can be used as support. The presence of glycogen in lipid cells has sometimes been described as detectable by P.A.S. with Schiff's reagent. Stains to detect mucin may be useful in case of myxoid differentiation. The fatty component of liposarcomas is Vimentin and S-100 positive obviously more evident in the well-differentiated forms. Well-differentiated liposarcoma stains also for MDM2 (inhibitor of p53, the "guardian of genoma") and CDK4 (Cyclin-dependent kinase 4).

Many liposarcomas can involve bone but often as a metastatic disease and not as primitive tumor; therefore, the absence of clinical evaluation of each single suspect case is fundamental to support the definitive diagnosis (Fig. 2.61).

Differential diagnosis includes lipoma and undifferentiated fibrosarcoma of bone (Fig. 2.62).

Liposarcoma

Definition: Liposarcoma of bone is a tumor of domestic animals with specific fat phenotype and that arises with adipose origin usually within the bone or in its surface.

Etiology: pre-existing lipoma.

Clinical presentations: painful large, soft-firm mass, with irregular and not well-defined margins, infiltrative and sometimes lytic and often myxoid.

Macroscopic features: osteolytic infiltrative bone lesion and usually is a high-grade bone tumor developing primarily in the long tubular bones.

Microscopic pathology: histopathology of liposarcoma of bone is similar to those of soft tissues.

Differential diagnosis: lipoma and undifferentiated fibrosarcoma of bone.

2.12.11 Malignant Mesenchymoma

Malignant mesenchymoma represents a very rare malignant tumor characterized by several cellular types. This type of tumor consists of mesenchymal cells that differentiate into two or more malignant histotypes, giving rise to potential neoplastic combinations which may include osteosarcoma, liposarcoma, fibrosarcoma, chondrosarcoma, etc. Obviously the general and particular descriptions of these

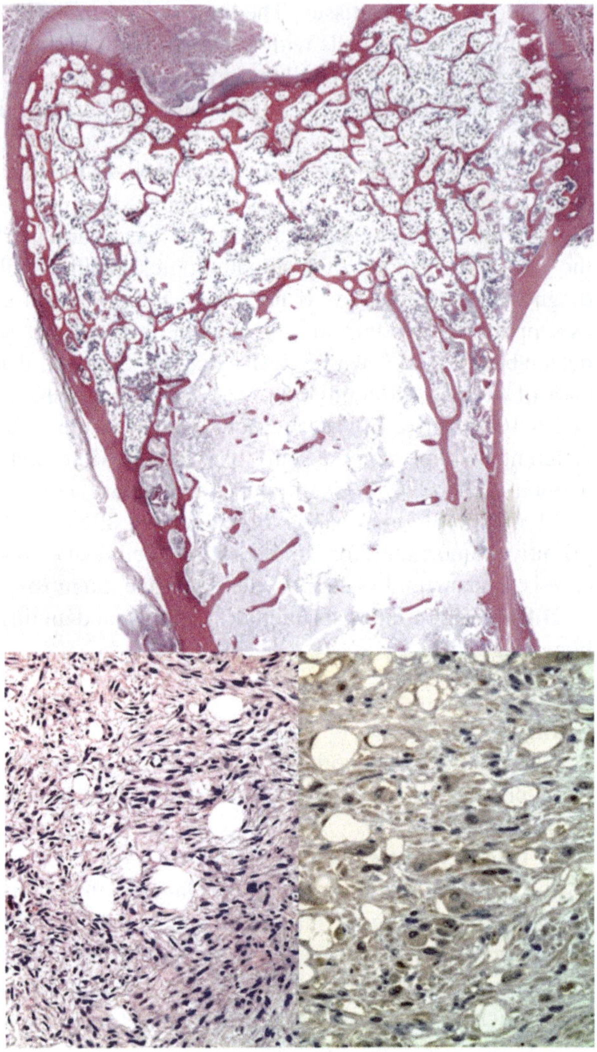

Fig. 2.61 Images of a liposarcoma that also may be considered by some as a pleomorphic liposarcoma in the most malignant region of this tumor of the proximal humerus of a 3-year-old neutered female domestic hair cat. Lesion was recognized after the cat fractured that bone. One histopathology image is stained with H&E stain. Tumor cells were positive for Sudan-O a lipid stain and is included the image stained with CD100 stain that was the recommended immunohistochemical stain for liposarcoma. (Thanks to Roy R. Pool, DVM, PhD Emeritus Professor of Pathology UCVD and Texas A&M—USA)

rare forms of tumor will be related to the cellular predominance and their morphological characteristics. In humans species a very rare main form of low-grade malignancy mesenchymoma is classified and called fibrocartilaginous mesenchymoma composed mainly of two types of tissues, one of which is benign cartilaginous and another with low-grade malignancy of the fibrosarcomatous type. It is mainly localized in different long bones and described also in ribs, metatarsus and arising as a slow growing scarcely painful solid, whitish dense mass (Figs. 2.63 and 2.64).

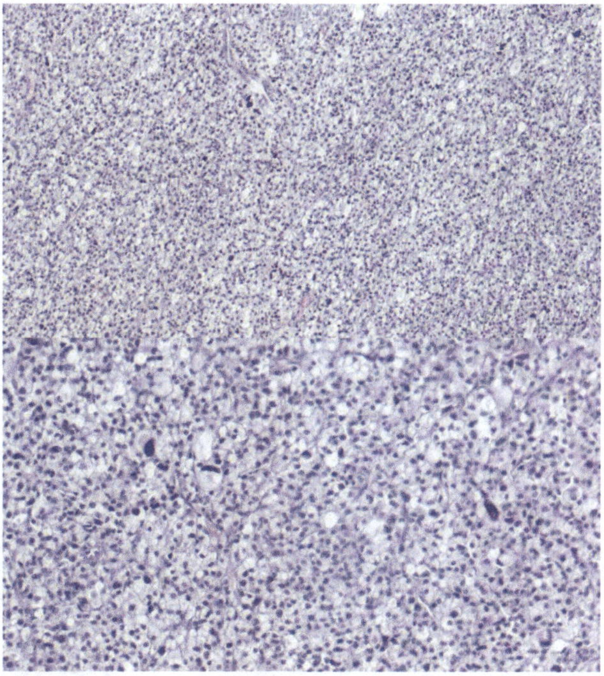

Fig. 2.62 Liposarcoma not well differentiated, dog, male, uncertain old (adult). (H&E, from author collection)

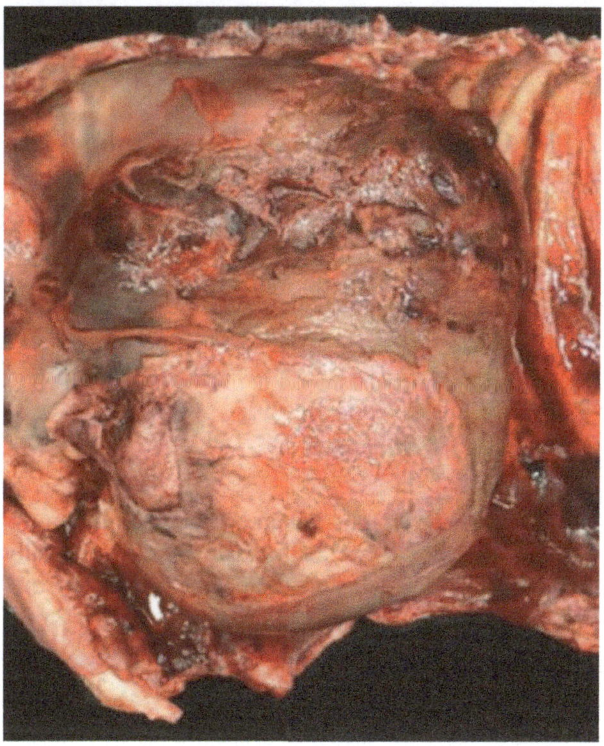

Fig. 2.63 Malignant mesenchymoma in the intercostal musculature of the rib case of a 2-year-old dog (Courtesy of Dr. Roy R. Pool, DVM, PhD, Emeritus Professor of Pathology UCVD and Texas A&M—USA)

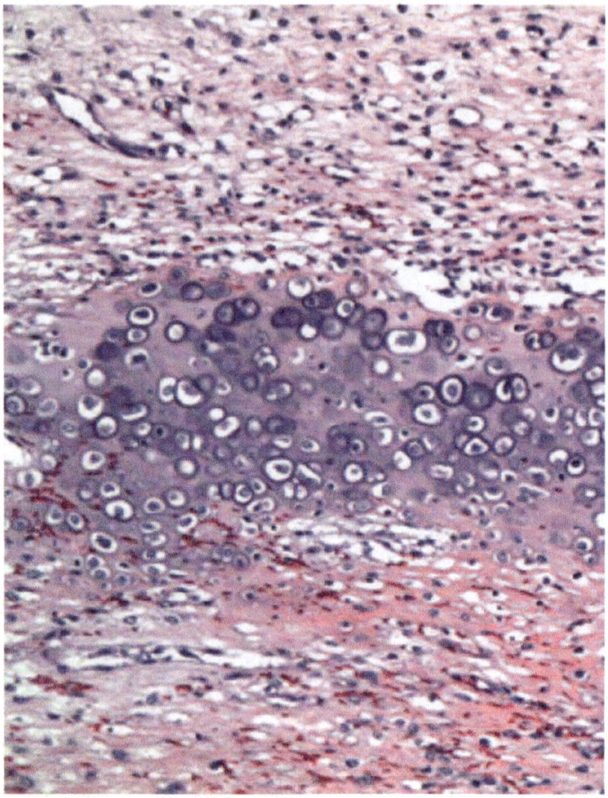

Fig. 2.64 Histologic figure from the malignant mesenchymoma in a site in which cartilage and early osteoid formation arises in a field of fibromyxoid lipoma (Courtesy of Dr. Roy R. Pool, DVM, PhD, Emeritus Professor of Pathology UCVD and Texas A&M—USA)

2.13 Others Uncommon Bone Tumors

There are other rare forms of bone tumors classified as "others" that include myxosarcoma as reported in WHO classification that occur mainly, also in its benign form of myxoma, in single joints of dogs as well reported in the book "Meuten, tumors of domestic animals."

2.13.1 Myxosarcoma

Myxosarcoma is viscous-mucinous producing tumor probably arising from joint's synoviocytes type B not distinguished radiographically from other malignant joint tumors. The general histopathological features of myxosarcoma are those of the predominant types of tumor cells (Figs. 2.65 and 2.66).

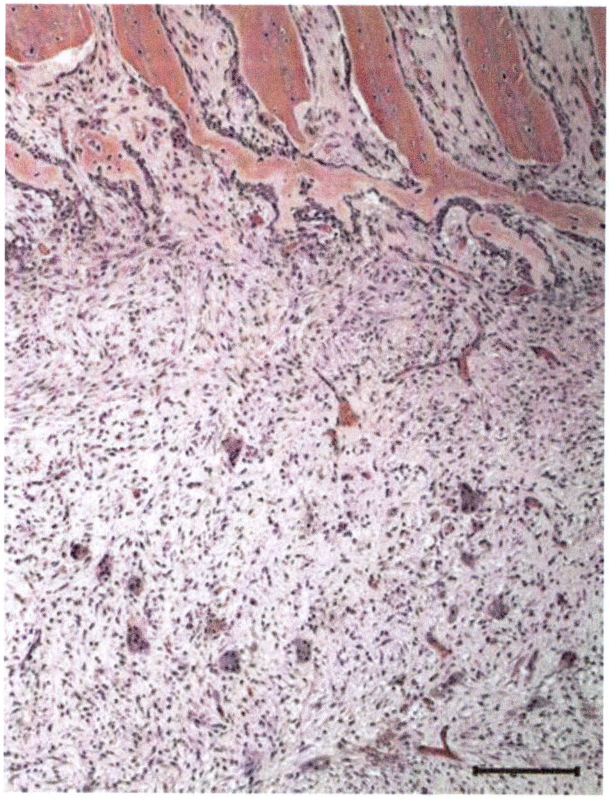

Fig. 2.65 Myxosarcoma from mandible of a dog (Courtesy of Dr. Roy R. Pool, DVM, PhD Emeritus Professor of Pathology UCVD and Texas A&M—USA)

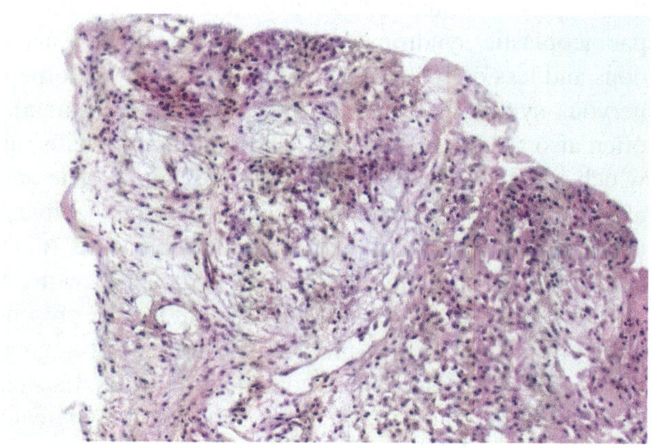

Fig. 2.66 Carpal low-grade sarcoma compatible with myxosarcoma and referred as a bone specimen (biopsy) from Veterinary practitioner and diagnosed definitively as sarcoma of soft tissues with low grade of malignancy. Dog, Labrador, 11 years old, male (H&E, from author collection), 10x

2.14 Tumors of Bone Marrow

2.14.1 Myeloma and Multiple Myeloma (MM)

Myeloma is a malignant tumor arising inside the bone marrow and composed predominantly of plasma cells (plasma cell myeloma [PCM]) that in human is the most common primary malignant bone tumor. In animals it has been reported in dogs, cats, and horses described also as a multiple myeloma where plasma cells play a fundamental role in the production of antibodies (immunoglobulins). It is precisely the disordered production of immunoglobulins that characterizes multiple myelomas which are also characterized by an evident monoclonal gammopathy which in dogs is expressed by IgG and/or IgA and in felines by IgG, and more rarely by light chains (myeloma of Bence-Jones) or heavy chains. Multiple myeloma in animals is also classified with a variant called non-secreting where the monoclonal gammopathy is absent at diagnosis and where the M protein (also called Spike protein or paraprotein) is also missing. This fragment of abnormal immunoglobulin, instead, is produced in excess in the multiple myeloma from an abnormal monoclonal proliferation of plasma cells. This tumor is rare in dogs and mainly affects purebred dogs, with an average age of 8–9 years, while in cats it is even rarer and in horses range near 10 years, from a study of only 10 cases. In the hypothesized pathogenetic mechanisms the genetic predisposition, deficit or diseases from strong and prolonged stimulation of the immune system, the direct action of some viruses are described. It appears that males are slightly more affected than females, but the epidemiological difference appears to be negligible.

From a clinical point of view, the picture can be variable in symptomatic manifestations, also depending on the location of the tumor, which include involvement and alterations of the bone tissue, hypercalcemia, hemorrhages, renal failure, heart failure, immunodeficiency and the so-called hyperviscous syndrome, a paraneoplastic syndrome, (involving not all animals with MM and only 20% in dogs and less of 10% in cats). Hyperviscous syndrome is clinically characterized by nervous symptoms (with seizures, nystagmus, ataxia), hemorrhagic retinopathies, often also associated with retinal detachment, uveitis, and increased ocular pressure which can also lead to blindness of the animal. The animal with multiple myeloma also frequently exhibits signs of arrhythmic tachycardia, often associated with cardio-respiratory insufficiency, which can lead to the formation of pulmonary edema. Another clinical finding associated with MM neoplastic disease is represented by coagulopathy responsible for the presence of systemic hemorrhages and epistaxis and due to the decrease in platelet aggregating factors and the inactivation of some plasma blood coagulation factors. Especially in dogs the Bence-Jones proteinuria associated with renal metastasis, hypercalcemia, dehydration, and the consequent reduction of renal filtration phenomena, represents adjuvant and predisposing factors for the condition of renal insufficiency detectable in the course of multiple myeloma. In bone localized and solitary forms of MM are diagnosed only occasionally and mainly in the canine and human species. Instead, polyostotic forms and the most frequent sites of bone involvement are represented by localizations in the vertebrae (especially in the thoracolumbar region), ribs, femur, pelvis, humerus, skull, and metaphysis of the long bones where, often, pathological fractures can also occur associated with myelopathy. Finally, it is important to point out the clinical data of hypercalcemia caused by the activation of osteoclastic stimulation factors by neoplastic cells which can also simultaneously form bone lysis and softening of bone. The presence of this condition of hypercalcemia can also be co-responsible for nervous symptoms, cardiac arrhythmias, and nephrosis phenomena. These clinical data are also associated with weight loss of the animal, lethargy, lameness, paraplegia due to the spinal cord compression when the tumor involves also the vertebrae.

Radiographic findings are characterized by lytic lesions centered in bone cavity, that sometimes showing multilobulated appearance. Usually when located in long bones the tumor is well circumscribed and may be surrounded by periosteal reaction with new bone formation. Solitary myeloma is also usually lytic with clear expansion in the bone. CT and MRI can show also thin lesions not detectable with radiographs.

Macroscopically the tumor appears as friable soft mass, of red-brownish color. Occasionally myeloma may metastasize to regional lymph nodes.

Microscopic features of myeloma can show several patterns. **Cytologic features** of well-differentiated myelomas reveal that smears contain numerous single and non-cohesive round to polygonal plasmacytoid cells with central round nuclei, with uniform size and with a typical distribution of chromatin also defined "spoke wheel," associated with prominent nucleoli. These cells show basophilic cytoplasm in abundant amounts with achromatic perinuclear areas corresponding to the Golgi apparatus. Smear from myeloma can show also binucleated and trinucleated cells also associated with multinucleated anaplastic giant cells, especially in the not well-differentiated forms. Mitotic figures are frequently detectable and associated with different degrees of cellular atypia and increased ratio of nucleus to cytoplasm. **Histopathology** can show different patterns, but generally myeloma is characterized by multinodular or sheets of densely packed plasma cells with a large spectrum of morphological features associated with normal bone marrows elements. In myelomas plasma cells can vary and show different morphological patterns that vary from mature plasma cells, to pleomorphic plasma cells especially in the not well-differentiated forms. The morphological pattern of neoplastic cells varies and when the tumor is well differentiated frequently they also do not appear as neoplastic. Mature plasma cells have roundish shape with basophilic cytoplasm and eccentric nuclei with "spoke wheels" chromatin. Distribution of chromatin can also have a different pattern called "clock-face" with more homogenous regular distribution of it. Associated to mature plasma cells is possible to detect plasmablasts with have dispersed nuclear chromatin and more prominent nucleoli. The undifferentiated plasma cells instead are pleomorphic and may contain multiple bluish cytoplasmic inclusions (Mott or Morula cells), Russell bodies (round reddish bodies) due to single or multiple cytoplasmic globules of immunoglobulins (Ig), fibrils, and crystals pathognomonic for diagnosis of myeloma.

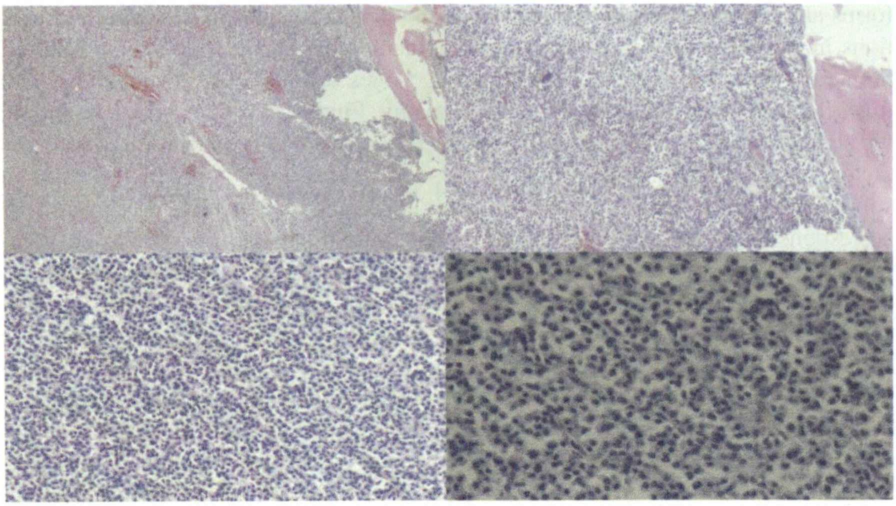

Fig. 2.67 Myeloma in a dog. Multinodular or sheets of densely packed plasma cells with a large spectrum of morphological features (from author collection)

Immunohistochemistry can be an important diagnostic support as different markers can be used for the definitive diagnosis of Myeloma. CD138, CD38, CD20, CD79a, and MUM1 are markers to detect plasma cells with CD19 always negative (Fig. 2.67).

2.14.2 Lymphoma

Primary malignant lymphoma of bone is a rare tumor described in dogs, cats, and cattle with frequent primary localizations in the long bones and in bone marrow. Malignant lymphoma of bone occurs also in humans where is also called non-Hodgkin lymphoma of bone and is still of uncertain origin. In animals often malignant lymphoma of bone has features of large, lytic, and destructive lesion that can involve cortex in a permeative growth often also associated to periosteal reaction. Epidemiological data reports that the tumor can occur also in young dogs sometime less of 1 year of age and involve both appendicular and axial skeleton where may cause pathologic fracture. Some authors report for canine malignant lymphoma of bone also a sporadic condition of hypercalcemia. When localized in vertebrae the neoplastic disease may be associated to neurological deficits for nerve impingement.

Radiographs show permeative and destructive pattern, with not well clear margins and foci of sclerosis to the affected bone.

Cytologically smears from aspirates show a uniform pattern of high or moderate amount of round large monomorphous lymphoid cells with scant or moderate basophilic and frequently vacuolated cytoplasm in a background with red blood cells. These cells have incised and non-incised nuclei, with prominent single or multiple nucleoli.

Histology of involved bone by malignant lymphoma is characterized by B or T phenotypical cells in the bone marrow cavity and in the involved other bone sections (cortex, trabeculae, etc.). Meuten report infiltration of bone marrow associated with infarction of bone in calves affected of bone's sporadic lymphoma, where the tumor spread also in the liver, kidney, and spleen.

Immunohistochemistry is always very helpful to classify malignant lymphoma of bone. CD30, CD25, CD45, and EMA show different variability of positive in large cell lymphoma, while CD68, CD3, and CD15 are usually negative.

Bibliography

Adler C-P (2000) Bone diseases: macroscopic, histological, and radiological diagnosis of structural changes in the skeleton, 1st edn. Springer, London

Aeffner F, Weeren R et al (2012) Synovial osteochondromatosis with malignant transformation to chondrosarcoma in a dog. Vet Pathol 49:1036–1039

Ahmed AR, Tan T-S et al (2003) Secondary chondrosarcoma in osteochondroma: report of 107 patients. Clin Orthop 411:193–206

Alhousami T, Sabharwal A, Gupta S, Aguirre A, Park E, Kramer JM (2018) Fibromyxoma of the jaw: case report and review of the literature. Head Neck Pathol 12:44–51. https://doi.org/10.1007/s12105-017-0823-0

American Registry of Pathology (ed) (n.d.) Joint tumors of domestic animals, vol 1, 2nd edn. American Registry of Pathology, Washington D.C

Atkins GJ, Kostakis P, Vincent C, Farrugia AN, Houchins JP, Findlay DM, Evdokiou A, Zannettino AC (2006) RANK expression as a cell surface marker of human osteoclast precursors in peripheral blood, bone marrow, and giant cell tumors of bone. J Bone Miner Res 21: 1339–1349

Avallone G, Rasotto R, Chambers JK, Miller AD, Behling-Kelly E, Monti P, Berlato D, Valenti P, Roccabianca P (2021) Review of histological grading Systems in Veterinary Medicine. Vet Pathol 58(5):809–828. https://doi.org/10.1177/0300985821999831

Baez JL, Hendrick MJ et al (2004) Liposarcomas in dogs: 56 cases (1989–2000). J Am Vet Med Assoc 224:887–891

Banks TA, Straw RC (2004) Multilobular osteochondrosarcoma of the hard palate in a dog. Aust Vet J 82:409–412

Berg J, Gliatto JM et al (1990) Giant cell tumor of the accessory carpal bone in a dog. J Am Vet Med Assoc 197:883–885

Berzina I, Sharkey LC et al (2008) Correlation between cytologic and histopathologic diagnoses of bone lesions in dogs: a study of the diagnostic accuracy of bone cytology. Vet Clin Pathol 37: 332–338

Bhandarkar GP, Shetty KV, Desai D (2020) Intramandibular plexiform schwannoma presenting as a nonspecific benign lesion: how specific can we be? A case report with a review of the literature. J Cancer Res Ther 16(3):641–646. https://doi.org/10.4103/jcrt.JCRT_393_19

Bhatti S, Ham LV et al (2001) Atlantoaxial cartilaginous exostosis causing cord compression in a mature Bernese mountain dog. J Small Anim Pract 42:79–81

Bienzle D, Silverstein DC et al (2000) Multiple myeloma in cats: variable presentation with different immunoglobulin isotypes in two cats. Vet Pathol 37:364–369
Binanti D, De Zani D, Fantinato E, Allevi G, Sironi G, Zani DD (2015) Intradural-extramedullary haemangioblastoma with paraspinal extension in a dog. Aust Vet J 93(12):460–465
Blair WH, Levine JM, Kerwin SC et al (2011) Imaging diagnosis – synovial myxoma of lumbar vertebrae articular process joint. Veterinary Radiol Ultrasound 52:309–312
Brewer WG, Turrel JM (1982) Radiotherapy and hyperthermia in the treatment of fibrosarcomas in the dog. J Am Vet Med Assoc 181:146–150
Bridge JA, Neff JR, Bhatia PS, Sanger WG, Murphey MD (1990) Cytogenetic findings and biologic behavior of giant cell tumors of bone. Cancer 65:2697–2703
Britt LG, Middleton OR et al (1998) Facial fibrosarcoma in two cows. Vet Radiol Ultrasound 39: 18–21
Brodey RS, Riser WH (1966) Liposarcoma of bone in a dog: a case report. Vet Radiol 7:27–33
Brodey RS, Sauer RM et al (1963) Canine bone neoplasms. J Am Vet Med Assoc 143:471–495
Brookes M (1972) Blood supply of developing bone and its possible bearing on malformation of the limbs and face in congenital haemangiomatous disorders. Proc R Soc Med 65(7):597–599
Brown RJ, Trevethan WP et al (1972) Multiple osteochondroma in a Siamese cat. J Am Vet Med Assoc 160:433–435
Brown KT, Kattapuram SV, Rosenthal DI (1986) Computed tomography analysis of bone tumors: patterns of cortical destruction and soft tissue extension. Skelet Radiol 15:448–451
Butch VL, Caron KM (2005) Blood and lymphatic vessel function. Cold Spring Harb Perspect Biol 7(3):1
Campanacci M (1999) Bone and soft tissue tumours: clinical features, imaging, pathology and treatment, 2nd edn. Springer-Verlag, Vienna
Campanacci M, Baldini N, Boriani S, Sudanese A (1987) Giant cell tumor of bone. The Journal of Bone & Joint Surgery 69:106–114
Cao L, Wu Y, Han SM, Sun T, Yu BH, Gao F, Wu WJ, Gao BL (2021) Dedifferentiated chondrosarcoma: a clinicopathologic analysis of 25 cases. BMC Musculoskelet Disord 22(1): 189. https://doi.org/10.1186/s12891-021-04053-7
Caporn TM, Read RA (1996) Osteochondromatosis of the cervical spine causing compressive myelopathy in a dog. J Small Anim Pract 37:133–137
Carrig CB, Seawright AA (1969) A familial canine polyostotic fibrous dysplasia with subperiosteal cortical defects. J Small Anim Pract 10:397–405
Ciekot PA, Powers BE et al (1994) Histologically low-grade, yet biologically high-grade, fibrosarcomas of the mandible and maxilla in dogs: 25 cases (1982–1991). J Am Vet Med Assoc 204: 610–615
Conti A, Caballero RG, Chiechi A, Dégano Blazquez RM, Barbado V, Krènacs T, Novello C, Pazzaglia L, Quattrini I, Picci P, De Alaya E, Benassi MS (2011) Identification of potential biomarkers for Giant cell tumor of bone using comparative proteomics analysis. Am J Pathol 178:88–97
Cook J, Turk JR et al (1998) Fibrosarcoma in the distal radius and carpus of a four-year-old Persian. J Am Anim Hosp Assoc 34:31–33
Cooley DM, Waters DJ (1997) Skeletal neoplasms of small dogs: a retrospective study and literature review. J Am Anim Hosp Assoc 33:11–23
Cornelius CE, Goodbary RF et al (1959) Plasma cell myelomatosis in a horse. Cornell Vet 49:478–493
Crow SE (1977) Primary hemangiosarcoma of the femur in a dog. Mod Vet Pract 58:343–346
Dahlin DC, Unni KK (1977) Osteosarcoma of bone and its important recognizable varieties. Am J Surg Pathol 1:61–72
Davis PE, Dixon RT et al (1974) Multiple liposarcoma of bone marrow origin in a greyhound. J Small Anim Pract 15:445–456
De Brot S, Grau-Roma L et al (2013) Occurrence of osteochondromatosis (multiple cartilaginous exostoses) in a domestic pig (Sus scrofa domesticus). J Vet Diagn Investig 25:599–602

De Vico G, Martano M, Maiolino P, Carella F, Leonardi L (2020) Expression of transferrin receptor-1 (TFR-1) in canine osteosarcomas. Vet Med Sci 6:272–276. https://doi.org/10.1002/vms3.258

Delisle F, Devauchelle P (1997) Gli Osteosarcomi nel cane. Summa 6:55–58

Dernell W, Straw R et al (1998) Multilobular osteochondrosarcoma in 39 dogs: 1979–1993. J Am Anim Hosp Assoc 34:11–18

Dernell WS, Ehrhart NP et al (2006) Tumors of the skeletal system. In: Withrow SJ, Vail DM (eds) Small Animal Clinical Oncology, 4th edn. Saunders Elsevier, Philadelphia, pp 540–582

Diamond SS, Raflo CP et al (1980) Multilobular osteosarcoma in the dog. Vet Pathol 17:759–763

Diaz-Bertrana C, Durall I et al (2010) Extra-and intra-articular synovial chondromatosis and malignant transformation to chondrosarcoma. Vet Comp Orthop Traumatol 23:277–283

Doige CE (1987) Multiple osteochondromas with evidence of malignant transformation in a cat. Vet Pathol 24(5):457–459. https://doi.org/10.1177/030098588702400515

Dorfman HD, Czerniak B (1998) Bone tumors. Mosby, St. Louis

Dorfman HD, Weiss SW (1984) Borderline osteoblastic tumors: problems in the differential diagnosis of aggressive osteoblastoma and low-grade osteosarcoma. Semin Diag Pathol 1: 215–234

Doussis I, Puddle B, Athanasou NA (1992) Immunophenotype of multinucleated and mononuclear cells in giant cell lesion of bone and soft tissue. J Clin Pathol 45:398–403

Dubielzig RR, Dickey DL (1978) Tracheal osteochondroma in a young dog. Vet Med Small Anim Clin 73:1288–1290

Durham AC, Popovitch CA et al (2008) Feline chondrosarcoma: a retrospective study of 67 cats (1987–2005). J Am Anim Hosp Assoc 44:124–130

Dürr H, Müller P et al (2002) Malignant lymphoma of bone. Arch Orthop Trauma Surg 122:10–16

Edwards DF, Parker JW et al (1993) Plasma cell myeloma in the horse. J Vet Intern Med 7:169–176

Enneking W.F. (1983). Musculoskeletal tumor surgery. Churchill Livingstone, Edinburgh, pp. 1–667

Farese JP, Kirpensteijn J et al (2009) Biologic behavior and clinical outcome of 25 dogs with canine appendicular chondrosarcoma treated by amputation: a veterinary society of surgical oncology retrospective study. Vet Surg 38:914–919

Ferrari A, Cozzi M, Aresu L, Martini V (2021) Tumor staging in a beagle dog with concomitant large B-cell lymphoma and T-cell acute lymphoblastic leukemia. J Vet Diagno Invest 33(4): 792–796. https://doi.org/10.1177/10406387211011024

Ferreras MC, Fuertes M, Pérez V, Benavides J, Garcìa-Parente C, Reyes LE, Garcìa-Marìn JF (2005) Giant cell tumour of bone in a cat with extraskeletal metastases: pathological and immunohistochemical study. J Vet Med A Physiol Pathol Clin Med 52(5):225–229

Fletcher CDM, Bridge JA, Hogendoorn PCW, Mertens F (2013) WHO classification of Tumours of soft tissue and bone. International Agency for Research on Cancer, Lyon

Fox SM, Cooke MM et al (1995) Primary intraosseous fibrosarcoma in a cat. NZ Vet J 43:197–200

Frase R, Freytag M et al (2009) Metastasizing liposarcoma of bone in a young dog. Vet Rec 164: 372–373

Frazier SA, Johns SM et al (2012) Outcome in dogs with surgically resected oral fibrosarcoma (1997–2008). Vet Comp Oncol 10:33–43

Gallegos J, Schwarz T, McAnulty JF (2008) Massive midline occipitotemporal resection of the skull for treatment of multilobular osteochondorsarcoma in two dogs. J Am Vet Med Assoc 233: 752–757

Garman RH, Powell FR et al (1977) Malignant giant cell tumor in a dog. J Am Vet Med Assoc 171: 546–548

Ghert M, Simunovic N et al (2007) Properties of the stromal cell in giant cell tumor of bone. Clin Orthop 459:8–13

Giebels F, Forterre F, Vincenti S, Geissbuehler U, Welle MM, Pool R, Soldati S, Maiolini A (2020) A complex histopathological challenge: suspicion of an osteoblastoma-like osteosarcoma

arising from the second thoracic vertebra in a cat. BMC Vet Res 16(1):378. https://doi.org/10.1186/s12917-020-02596-w
Giger U, Evans SM et al (1989) Orthovoltage radiotherapy of primary lymphoma of bone in a dog. J Am Vet Med Assoc 195:627–630
Ginaldi S, Williams CD (1985) Seeding of malignant lymphoma along the tract after bone marrow biopsy. South Med J 78:1007–1008
Goedegebuure SA, Firth EC, Dik KJ (1983) Osteoblastoma in the radius of a pony. Vet Pathol 20: 650–652
Goldring SR, Dayer JM, Russell RG, Mankin HJ, Krane SM (1978) Response to hormones of cells cultured from human giant cell tumors of bone. J Clin Endocrinol and Metabol 46:425–433
Goldring SR, Schiller AL, Mankin HJ, Dayer JM, Krane SM (1986) Characterization of cells from human giant cell tumors of bone. Clin Orthop 204:59–75
Goldschmidt MH, Hendrick MJ (2002) Tumors of the skin and soft tissues. In: Meuten DJ (ed) Tumors in domestic animals, 4th edn. Iowa State Press, Ames, pp 45–117
Gómez-Laguna J, Barranco I, Rodríguez-Gómez IM, Blanco B, Guil-Luna S, Carrasco L, de las Mulas JM. (2012) Malignant mesenchymoma of the heart base in a dog with infiltration of the pericardium and metastasis to the lung. J Comp Pathol 147(2–3):195/8. https://doi.org/10.1016/j.jcpa.2012.01.002
Gourley IMG, Morgan JP et al (1970) Tracheal osteochondroma in a dog: a case report. J Small Anim Pract 11:327–335
Grano M, Colucci S, De Bellis M, Zigrino P, Argentino L, Zambonin G, Serra M, Scotlandi K, Teti A, Zambonin ZA (1994) New model for bone resorption study in vitro: human osteoclast-like cells from giant cell tumors of bone. J Bone and Min Res 9:1013–1020
Green EM, Adams WM et al (1999) Malignant transformation of solitary spinal osteochondroma in two mature dogs. Vet Radiol Ultrasound 40:634–637
Grundmann E (1976) Recent results in cancer research – malignant bone tumors. Springer, Berlin
Guard C, Wilkinson JE (1984) Hemangiosarcoma in a cow. J Am Vet Med Assoc 185:789–790
Hahn KA, McGavin MD et al (1997) Bilateral renal metastases of nasal chondrosarcoma in a dog. Vet Pathol 34:352–355
Hammer AS, Weeren FR et al (1995) Prognostic factors in dogs with osteosarcomas of the flat or irregular bones. J Am Anim Hosp Assoc 31:321–326
Haque AU, Moatasim A (2008) Giant cell tumor of bone: a neoplasm or a reactive condition? Int J Clin Exp Pathol 1:489–501
Harris RA, Miller M, Donaghy D, Ashton L, Langston C, Shockey T, Moore AR (2021) Light chain myeloma and detection of free light chains in serum and urine of dogs and cats. J Vet Intern Med 35(2):1031–1040. https://doi.org/10.1111/jvim.16070
Hathcock JT, Newton JC (2000) Computed tomographic characteristics of multilobular tumor of bone involving the cranium in 7 dogs and zygomatic arch in 2 dogs. Vet Radiol Ultrasound 41: 214–217
Hay CW, Roberts R et al (1994) Multilobular tumour of bone at an unusual location in the axilla of a dog. J Small Anim Pract 35:633–636
Heck RK, Jr Canale ST, Beaty JH (2008) Campbell's operative orthopedics. Malignant bone tumors. Mosby Year Book, Missouri, pp 901–938
Henriksen MDL, Sharkey L, Wunschmann A, Franzen-Klein D, Willette M, Teixeira LBC, Dubielzig R, Henriksen MDL (2019) Bilateral anterior uveitis in a northern saw-whet owl (aegolius acadicus) with a metastatic pectoral malignant Mesenchymoma. J Avian Med Surg 33(2):171–178. https://doi.org/10.1647/2017-326
Hodik V, Loeb E et al (2012) Chondroma of the vertical ramus of the feline mandible. J Feline Med Surg 14:924–927
Inwards CY, Unni KK (1995) Classification and grading of bone sarcomas. Hematol Oncol Clin North Am 9:545–569
Iwata Y, Kondo H, Koizumi I, Shibuya H (2021) Tracheal chondrosarcoma in a blue tegu (Salvator merianae). J Vet Med Sci 83(6):981–983. https://doi.org/10.1292/jvms.21-0145

Jacobsen SA (1971) In: Thomas CC (ed) The comparative pathology of the tumors of bone. C.C. Thomas, cop, Springfield, IL

Jaffe HL (1958) Tumours and tumorous conditions of the bones and joints. Lea & Febiger, Philadelphia, p 629

Jorgensen JS, Geoly FJ et al (1997) Lameness and pleural effusion associated with an aggressive fibrosarcoma in a horse. J Am Vet Med Assoc 210:1328–1331

Joseph CG, Hwang H, Jiao Y, Wood LD, Kinde I, Wu J, Mandahl N, Luo J, Hruban RH, Diaz LA et al (2014) Exomic analysis of myxoid liposarcomas, synovial sarcomas, and osteosarcomas. Genes Chromosomes Cancer 53:15–24

Jubb KVF, Kennedy PC, Palmer N (2007) Pathology of domestic animals: chapter 1 – bone and joints, vol 1, 5th edn. Elsevier Saunders, New York, pp 119–120

Kang JY, Kim HJ, Wojno TH, Yeung AM, Mendoza PR, Grossniklaus HE (2021) Atypical lipomatous tumor/well-differentiated Liposarcoma of the orbit: three cases and review of the literature. Ophthalmic Plast Reconstr Surg 37(3S):134–140. https://doi.org/10.1097/IOP.0000000000001804

Kawakami T, Shimizu T, Kimura A, Hasegawa H, Siar CH, Ng KH, Nagatsuka H, Nagai N, Kanda H (2005) Immunohistochemical examination of cytological differentiation in osteosarcomas. Eur J Med Res 10:475–479

Kaya M, Wada T, Akatsuka T et al (2000) Vascular endothelial growth factor expression in untreated osteosarcoma is predictive of pulmonary metastasis and poor prognosis. Clin Cancer Res 6(2):572–577

Keel K, Bryan J et al (2009) Cranial osteochondromas in free-ranging whitetailed deer (Odocoileus virginianus). Vet Pathol 46:1065

Kenney PJ, Gilula LA, Murphy WA (1981) The use of computed tomography to distinguish osteochondroma and chondrosarcoma. Radiology 139:129–137

Kent MS, Strom A et al (2004) Alternating carboplatin and doxorubicin as adjunctive chemotherapy to amputation or limb-sparing surgery in the treatment of appendicular osteosarcoma in dogs. J Vet Intern Med 18:540–544

Kim H, Nakaichi M, Itamoto K, Taura Y (2007) Primary chondrosarcoma in the skull of a dog. J Vet Sci 8(1):99–101. https://doi.org/10.4142/jvs.2007.8.1.99

Kim H, Lee YS, Kim KH, Kim JY, Yoo SJ et al (2017) Aggressive osteoblastoma of the cervical vertebrae in a Turkish angora cat. Turk J Vet Anim Sci 41:444–448

Kinpara K (2000) Osteoclast differentiation factor in human osteosarcoma cell line. J Immunoass 21(4):327–340

Kirk NM, Vieson MD, Sullivan DR, Pool RR (2019) Osteoblastoma in the distal humerus of a cat. J Comp Pathol 172:27–30

Kirpensteijn J, Kik M et al (2002) Prognostic significance of a new histologic grading system for canine osteosarcoma. Vet Pathol 39:240–246

Kirpensteijn J, Kik M, Teske E, Rutteman GR (2008) TP53 gene mutations in canine osteosarcoma. Vet Surg 37:454–460

Klein MJ, Siegal GP (2006) Osteosarcoma: anatomic and histologic variants. Am J Clin Pathol 125: 555–581

Klenke FM, Wenger DE, Inwards CY, Rose PS, Sim FH (2010) Giant cell tumor of bone, risk factors for recurrence. Clin Orthop Relat Res 469:591–599

Knecht CD, Priester WA (1978) Musculoskeletal tumors in dogs. J Am Vet Med Assoc 172:72–74

Kpshy M, Paulino AC, Mai WY, Teh BS (2005) Radiation-induced osteosarcomas in the pediatric population. Int J Radiat Oncol Biol Phys 63:1169–1174

Kreuter M, Bieker R, Bielaek SS et al (2004) Prognostic relevance of increased angiogenesis in osteosarcoma. Clin Cancer Res 10(24):8531–8537

Kumta SM, Huang L, Cheng YY, Chow LT, Lee KM, Zheng MH (2003) Expression of VEGF and MMP-9 in giant cell tumor of bone and other osteolytic lesions. Life Sci 73:1427–1436

Kundu ZS (2014) Classification, imaging, biopsy and staging of osteosarcoma. Indian J Orthop 48(3):238–246

Kuntz C, Dernell W et al (1998) Extraskeletal osteosarcomas in dogs: 14 cases. J Am Anim Hosp Assoc 34:26–30

Lamagna B, Lamagna F et al (2006) Polyostotic lymphoma with vertebral involvement and spinal extradural compression in a dog. J Am Anim Hosp Assoc 42:71–76

Langenbach A, Anderson M et al (1998) Extraskeletal osteosarcomas in dogs: a retrospective study of 169 cases (1986–1996). J Am Anim Hosp Assoc 34:113–120

Langley Hobbs SJ, Carmichael S et al (1997) Polyostotic lymphoma in a young dog: a case report and literature review. J Small Anim Pract 38:412–416

LaRue SM, Withrow SJ et al (1986) Radiographic bone surveys in the evaluation of primary bone tumors in dogs. J Am Vet Med Assoc 188:514–516

Layfield LJ (2009) Cytologic diagnosis of osseous lesions: a review with emphasis on the diagnosis of primary neoplasms of bone. Diagn Cytopathol 37:299–310

Layfield LJ, Emerson L, Crim JR et al (2008) Squamous differentiation and cytokeratin expression in an osteosarcoma. A case report and review on the literature. Clin Med Pathol 1:55–59

LeCouteur RA, Nimmo JS et al (1978) A case of giant cell tumor of bone (osteoclastoma) in a dog. J Am Anim Hosp Assoc 14:356–362

Lee B-R, Lee S-H et al (2011) Abdominal aortic chondrosarcoma in a dog. J Vet Med Sci 73:1473–1476

Lemsawatdikul K, Gooding CA, Twomey EL, Kim GE, Goldsby RE, Cohen L, O'Donnell RJ (2005) Seeding of osteosarcoma in the biopsy tract of a patient with multifocal osteosarcoma. Pediatr Radiol 35:717–721

Leonardi L, Mechelli L, Bellezza E, Benassi MS (2012a) A comparative study on phenotypical and biomolecular characterization of giant cell tumor of bone in feline and human species. J Biomed Bioeng 3(1):71–78

Leonardi L et al (2012b) An unusual case report of primitive Jejuneal canine osteosarcoma. Int J Vet Sci 1(2):69–71

Leonardi L, Quattrini I, Roperto F, Benassi MS (2013) Protease expression in giant cell tumour of bone: a comparative study on feline and human samples. Res Vet Sci 95(2):310–315

Leonardi L, Carrano A, Stoppini L, Floris M (2014a) Multilobular tumor of the zygomatic bone in a dog. Open Vet J 4:9–11

Leonardi L et al (2014b) Fibroblastic osteosarcoma in a lion (Panthera leo). Open Vet J 4(1):1–3

Leonardi L, Pazzaglia L, Benassi MS (2018) miR-1 and miR-133b expression in canine osteosarcoma. Res Vet Sci 117:133–137. https://doi.org/10.1016/j.rvsc.2017.12.002

Leonardi L, Benaasi MS, Pollino S, Locaputo C, Pazzaglia L (2020) miR-106B-25 cluster eexpression: a comparative human and canine osteosarcoma study. Vet Rec Open 7:e000379. https://doi.org/10.1136/vetreco-2019-000379

Leonardi L, Scotlandi K, Pettinari I, Benassi MS, Porcellato I, Pazzaglia L (2021) MiRNAs in canine and human osteosarcoma: a highlight review on comparative biomolecular aspects. Cell 10:428. https://doi.org/10.3390/cells10020428

Levitin B, Aroch I et al (2003) Linear osteochondromatosis in a cat. Vet Radiol Ultrasound 44:660–664

Lichtenstein L (1972) Bone tumors, 4th edn. Mosby, St Louis

Liptak JM, Dernell WS et al (2004) Proximal radial and distal humeral osteosarcoma in 12 dogs. J Am Anim Hosp Assoc 40:461–467

Liu S, Dorfman HD, Patnaik AK (1974) Primary and secondary bone tumours in the cat. J Small Anim Pract 15:141–156

Liu SK, Dorfman HD et al (1977) Primary and secondary bone tumours in the dog. J Small Anim Pract 18:313–326

Liu B, Yu SF, Li TJ (2003) Multinucleated giant cells in various forms of giant cell containing lesions of the jaws express features of osteoclasts. J Oral Pathol Med 32:367–375

Livesay MA, Wilkie IW (1986) Focal and multifocal osteosarcoma in two foals. Equine Vet J 18: 407–410

Löhr CV, Hedge ZN, Pool RR (2012) Infiltrative myxoma of the stifle joint and thigh in a domestic rabbit (Oryctolagus cuniculus). J Comp Pathol 147(2–3):218–222
Lopez A (2011) Pathology of bones. http://people.upei.ca/lopez/bones/bone-notes.pdf
Losco PE, Diters RW, Walsh KM (1984) Canine multilobular osteosarcoma of the skull with metastasis. J Comp Pathol 94:621–624
Loukopoulos P, Robinson WF (2007) Clinicopathological relevance of tumour grading in canine osteosarcoma. J Comp Pathol 136:65–73
Lovell S, Burchell RK, Roady PJ, Fredrickson RL, Gal A (2017) BMC. Vet Res 13(1):247. https://doi.org/10.1186/s12917-017-1181-6
Lucroy MD, Clauson RM, Suckow MA, El-Tayyeb F, Kalinauskas A (2020) Evaluation of an autologous cancer vaccine for the treatment of metastatic canine hemangiosarcoma: a preliminary study. BMC Vet Res 16(1):447. https://doi.org/10.1186/s12917-020-02675
Maeda H, Ozaki K et al (1993) Bone lesions of multiple myeloma in three dogs. J Vet Med A Physiol Pathol Clin Med 40:384–392
Manor EK, Craig LE, Sun X, Cannon CM (2018) Prior joint disease is associated with increased risk of periarticular histiocytic sarcoma in dogs. Vet Comp Oncol 16(1):83–88. https://doi.org/10.1111/vco.12338
Manuali E, Forte C, Vichi G, Genovese DA, Mancini D, De Leo AAP, Cavicchioli L, Pierucci P, Zappulli V (2020) Tumours in European shorthair cats: a retrospective study of 680 cases. J Feline Med Surg 22(12):1095–1102. https://doi.org/10.1177/1098612X20905035
Marconato L, Del Piero F (2005) Oncologia Medica dei Piccoli Animali. Poletto Editore, Milano
Martinek B, Bago Z et al (2006) Chondrosarcoma in a Simmental cow: clinical, ultrasonographic, radiographic and pathological findings. Vet J 172:181–184
Mashima E, Sawada Y, Nakamura M (2021) Recent advancement in atypical Lipomatous tumor research. Int J Mol Sci 22(3):994. https://doi.org/10.3390/ijms22030994
Masserdotti C, Bonfanti U et al (2006) Use of oil red O stain in the cytologic diagnosis of canine liposarcoma. Vet Clin Pathol 35:37–41
Matus RE, Leifer CE et al (1986) Prognostic factors for multiple myeloma in the dog. J Am Vet Med Assoc 188:1288–1292
May SA, Baker JR (1985) A giant cell tumour (osteoclastoma) of the tibia in a horse. Equine Vet J 17:477–479
Mazzullo G, Russo M, Niutta PP, De Vico G (2004) Osteosarcoma with multiple metastases and subcutaneous involvement in a rabbit (*Oryctolagus cuniculus*). Vet Clin Pathol 33:102–104
McCalla TL, Moore CP et al (1989) Multilobular osteosarcoma of the mandible and orbit in a dog. Vet Pathol 26:92–94
McGavin MD, Zachary JF (2007) Pathologic basis of veterinary disease, 4th edn. Mosby-Elsevier, St. Louis, p 1091
McLain DL, Hill JR, Pulley LT (1983) Multilobular osteoma and chondroma (chondroma rodens) with pulmonary metastasis in a dog. J Am Anim Hosp Assoc 19:359–362
Mehl ML, Withrow SJ et al (2001) Spontaneous regression of osteosarcoma in four dogs. J Am Vet Med Assoc 219:614–617
Meierhenry EF (1974) Metastatic liposarcoma with extensive osteolysis in the dog. J Am Anim Hosp Assoc 10:478–481
Mellanby RJ, Holloway A et al (2003) Primary chondrosarcoma in the pulmonary artery of a dog. Vet Radiol Ultrasound 44:315–321
Mendez SE, Sykes Crumplar SE, Durham AC (2020) Primary Hemangiosarcoma of the Falciform fat in seven dogs (2007-2015). J Am Anim Hosp Assoc 56(2):120–126
Mendoza S, Konishi T, Dernell WS, Withrow SJ, Miller CW (1998a) Status of the p53, Rb and MDM2 genes in canine osteosarcoma. Anticancer Res 18:4449–4453
Mendoza SM, Konishi T, Miller CW (1998b) Integration of SV40 in human osteosarcoma DNA. Oncogene 17(19):2457–2462
Meuten DJ (2020) Tumors in domestic animals, 5th edn. Iowa State University Press, Ames

Meyers B, Boy SC, Steenkamp G (2007) Diagnosis and management of odontogenic myxoma in a dog. J Vet Dent 24(3):166–171. https://doi.org/10.1177/089875640702400304
Miller CW, Also A, Won A, Tan M, Lampkin B, Koeffler HP (1996) Alterations of the p53, Rb and MDM2 genes in osteosarcoma. J Cancer Res Clin Oncol 122(9):559–565
Miller MA, Pool RR, Coolman BR (2007) Synovial hemangioma in the stifle joint of a dog. Vet Pathol 44(2):240–243
Mirra JM, Gold R, Downs J, Eckardt JJ (1985) A new histologic approach to the differentiation of Enchondroma and chondrosarcoma of the bones. A Clinicopathologic analysis of 51 cases. Clin Orthop Relat Res 201:214–237
Misdorp W, Hart AAM (1979) Some prognostic and epidemiologic factors in canine osteosarcoma. J Natl Cancer Inst 62:537–545
Miwa Y, Nakata M, Takimoto H, Chambers JK, Uchida K (2021) Spontaneous oral tumours in 18 rabbits (2005-2015). J Small Anim Pract 62(2):156–160. https://doi.org/10.1111/jsap.13082
Monti P, Barnes D, Adrian AM, Rasotto R (2018) Synovial cell sarcoma in a dog: a misnomer-cytologic and histologic findings and review of the literature. Vet Clin Pathol 47(2):181–185. https://doi.org/10.1111/vcp-12590.Epub2018
Moore PF (2014) A review of histiocytic diseases of dogs and cats. Vet Pathol 51(1):167–184
Morgan JP, Ackerman N et al (1980) Vertebral tumors in the dog: a clinical, radiologic, and pathologic study of 61 primary and secondary lesions. Vet Radiol 21:197–212
Morgan T, Atkins GJ, Trivett MK, Johnson SA, Kansara M, Schlicht SL, Slavin JL, Simmons P, Dickinson I, Powell G, Choong PF, Holloway AJ, Thomas DM (2005) Molecular profiling of giant cell tumor of bone and the osteoclastic localization of ligand for receptor activator of nuclear factor kappaB. Am J Pathol 167:117–128
Morton D (1985) Chondrosarcoma arising in a multilobular chondroma in a cat. J Am Vet Med Assoc 186:804–806
Moskaleva EY, Zhorov ES, Semochkina YP, Rodina AV, Vysotskaya OV, Glukhov AI, Chukalova AA, Posypanova GA, Saprykin VP (2017) Characteristics of tumors that developed in mice after treatment with irradiated syngeneic mesenchymal stem cells of bone marrow. Tsitologiia 59(4):271–278
Moss SF, Blser MJ (2005) Mechanisms of disease: inflammation and the origins of cancer. Nat Clin Pract Oncol 2:90–97
Mozos E, Novales M, Ginel PJ, Pérez J, Pool RR (2002) A newly recognized pattern of canine osteochondromatosis. Vet Radiol Ultrasound 43(2):132–137
Mueller F, Fuchs B, Kaser-Hotz B (2007) Comparative biology of human and canine osteosarcoma. Anticancer Res 27:155–164
Munday JS, Prahl A (2002) Retroperitoneal extraskeletal mesenchymal chondrosarcoma in a dog. J Vet Diagn Investig 14:498–500
Murphey MD, Arcara LK et al (2005) Imaging of musculoskeletal liposarcoma with radiologic-pathologic correlation. Radiographics 25:1371–1395
Nelson AM, Baker DC (1998) Pedal osteosarcoma in a donkey. Vet Pathol 35:407–409
Nielsen GP, Rosenberg AE (2012) Diagnostic pathology bone, 1st edn. Amirsys Inc., Philadelphia
Nolff MC, Puff C, Länger B, Fehr M (2014) Feline osteochondromatosis in a FELV-negative European shorthair cat. Tierarztl Prax Ausg K Kleintiere Heimtiere 42(1):55–59
Nye CJ, Mustonen AM, Cook L, Hostnik ET (2021) Bone hemangiosarcoma in the calvarium of a dog. J Small Anim Pract. https://doi.org/10.1111/jsap.13355
Okada K, Hasegawa T, Yokoyama R et al (2003) Osteosarcoma with cytokeratin expression: a clinicopathological study of six cases with an emphasis on differential diagnosis from metastatic cancer. J Clin Pathol 56:742–746
Osborne CA, Perman V et al (1968) Multiple myeloma in the dog. J Am Vet Med Assoc 153:1300–1319
Ottaviani G, Jaffe N (2009) The epidemiology of osteosarcoma. Cancer Treat Res 152:3–13
Pakhrin B, Bae IH et al (2006) Multilobular tumor of the mandible in a Pekingese dog. J Vet Sci 7: 297–298

Paoloni M, Davis S, Lana S, Withrow S, Sangiorgi L, Picci P, Hewitt S, Triche T, Meltzer P, Khanna C (2009) Canine tumor cross-species genomics uncovers targets linked to osteosarcoma progression. BMC Genomics 10:625

Patel RT, Caceres A et al (2005) Multiple myeloma in 16 cats: a retrospective study. Vet Clin Pathol 34:341–352

Patnaik AK (1990) Canine extraskeletal osteosarcoma and chondrosarcoma: a clinicopathologic study of 14 cases. Vet Pathol 27:46–55

Patnaik AK, Lieberman PH et al (1984) Canine sinonasal skeletal neoplasms: chondrosarcomas and osteosarcomas. Vet Pathol 21:475–482

Pazzaglia L, Leonardi L, Conti A, Novello C, Quattrini I, Montanini L, Roperto F, Del Piero F, Di Guardo G, Piro F, Picci P, Benassi MS (2015) miR196a expression in human and canine osteosarcomas: a comparative study. Res Vet Sci 99:112–119. https://doi.org/10.1016/j.rvsc.2014.12.017

Peiffer RL, Rebar A et al (1974) Fibrosarcoma involving the skeleton of the dog. Vet Med Small Anim Clin 69:1143–1148

Phillips JC, Lembcke L, Chamberlin T (2010) A novel locus for canine osteosarcoma (OSA1) maps to CFA34, the canine orthologue of human 3q26. Genomics 96:220–227

Picci P (2007) Osteosarcoma (osteogenic sarcoma). Orphanet J Rare Dis 2:1. article 6

Pool RR (1990) Tumors of bone and cartilage. In: Moulton JE (ed) Tumors in domestic animals, 3rd edn. University of California Press, Berkeley, pp 207–213

Pool RR, Carrig CB (1972) Multiple cartilaginous exostosis in a cat. Vet Pathol 9:350–359

Pool RR, Harris JM (1975) Feline osteochondromatosis. Feline Pract 5:24–30

Pool RR, Wolf HG (1974) An unusual case of canine osteosarcoma. Cancer 34(3):771–779

Pool RR, Morgan JP, Parks NJ, Farnham JE, Littman MS (1983) Comparative pathogenesis of radium-induced intracortical bone lesions in humans and beagles. Health Phys 44(Suppl 1): 155–157

Popovitch CA, Weinstein MJ et al (1994) Chondrosarcoma: a retrospective study of 97 dogs (1987–1990). J Am Anim Hosp Assoc 30:81–85

Popp JA, Simpson CF (1976) Feline malignant giant cell tumor of bone associated with C-type virus particles. Cornell Vet 66:528–535

Psychas V, Loukopoulos P, Polizopoulou ZS, Sofianidis G (2009) Multilobular tumour of the caudal cranium causing severe cerebral and cerebellar compression in a dog. J Vet Sci 10:81–83

Rahe MC, Westegaard T, Yaeger M (2020) Extraskeletal chondrosarcoma in the tongue of a dog: case report and retrospective analysis of 236 tongue masses (2011-2019). J Vet Diagn Investig 32(3):440–443. https://doi.org/10.1177/1040638720911394

Ramírez GA, Altimira J, Vilafranca M (2015) Cartilaginous tumors of the larynx and trachea in the dog: literature review and 10 additional cases (1995-2014). Vet Pathol 52(6):1019–1026. https://doi.org/10.1177/0300985815579997

Ramos-Vara JA, Miller MA, Valli VEO (2007) Immunohistochemical detection of multiple myeloma 1/interferon regulatory factor 4 (MUM1/IRF-4) in canine Plasmacytoma: comparison with CD79a and CD20. Vet Pathol 44:875–884

Ranade SA, Pacchiana PD (2011) What is your diagnosis? Osteochondromatosis. JAVMA 238(10): 1243–1244

Reinhardt S, Stockhaus C et al (2005) Assessment of cytological criteria for diagnosing osteosarcoma in dogs. J Small Anim Pract 46:65–70

Resnick D, Kyriakos M et al (1995) Tumors and tumor-like lesions of bone: imaging and pathology of specific lesions. In: Resnick D (ed) Diagnosis of bone and joint disorders, vol 6, 3rd edn. W.B. Saunders, Philadelphia, pp 3628–3938

Richardson DW, Acland HM (1983) Multilobular osteoma (chondroma rodens) in a horse. J Am Vet Med Assoc 182:289–291

Riddle WE, Leighton RL (1970) Osteochondromatosis in a cat. J Am Vet Med Assoc 156:1428–1430

Rigopoulou A, Saifuddin A (2012) Intraosseous hemangioma of the appendicular skeleton: imaging features of 15 cases, and a review of the literature. Skelet Radiol 41:1525–1536
Rodenas S, Pumarola M et al (2008) Imaging diagnosis – cervical spine chondroma in a dog. Vet Radiol Ultrasound 49:464–466
Roessner A, von Bassewitz DB, Schlake W, Thorwesten G, Grundmann E (1984) Biologic characterization of human bone tumors. III. Giant cell tumor of bone. A combined electron microscopical, histochemical, and autoradiographical study. Pathol Res Pract 178:431–440
Rolvien T, Zustin J, Amling M, Yastrebow O (2018) Periosteal chondroma of the cuboid with secondary aneurysmal bone cyst in a setting of secondary hyperparathyroidism. Foot Ankle Surg 24:71–75
Rosenberg AE (2010) Robins and Cotran pathologic basis of disease. Bone, joints and soft tissue tumors, 8th edn. WB Saunders, Philadelphia, pp 1203–1256
Ru G, Terracini B et al (1998) Host related risk factors for canine osteosarcoma. Vet J 156:31–39
Ryseff JK, Bohn AA (2012) Detection of alkaline phosphatase in canine cells previously stained with Wright–Giemsa and its utility in differentiating osteosarcoma from other mesenchymal tumors. Vet Clin Pathol 41:391–395
Salerno M, Avnet S, Alberghini M, Giunti A, Baldini N (2008) Histogenetic characterization of Giant cell tumor of bone. Clin Orthop Relat Res 466:2081–2091
Savage SA, Mirabello L (2011) Using epidemiology and genomics to understand osteosarcoma etiology. Sarcoma 2011:548151
Schajowicz F, Lemos C (1970) Osteoid osteoma and osteoblastoma. Closely related entities of osteoblastic derivation. Acta Orthop Scand 41:272–291
Schmid T, Hilbe M et al (2010) Chondrosarcoma in the humerus of a goat. Vet Comp Orthop Traumatol 23:273–276
Silver GM, Bagley RS et al (2001) Radiographic diagnosis: cartilaginous exostoses in a dog. Vet Radiol Ultrasound 42:231–234
Silverman J, Weisbrode SE et al (1994) Enchondroma in a rhesus monkey. J Am Vet Med Assoc 204:786–788
Simpson S, Dunning MD, de Brot S, Grau-Roma L, Mongan NP, Rutland CS (2017) Comparative review of human and canine osteosarcoma: morphology, epidemiology, prognosis, treatment and genetics. Acta Vet Scand 59:71
Slauson DO, Cooper BJ (2002) Mechanisms of disease a textbook of comparative general pathology, 3rd edn. Mosby, St. Louis
Slayter MV, Boosinger TR, Pool RR, Dämmrich K, Misdorp W, Larsen S (1994) Histological classification of bone and joint tumors of domestic animals, Second series, vol 1. AFIP, Washington DC
Spodnick GJ, Berg J et al (1992) Prognosis for dogs with appendicular osteosarcoma treated by amputation alone: 162 cases (1978–1988). J Am Vet Med Assoc 200:995–999
Spoldi E, Schwarz T, Sabattini S, Vignoli M, Cancedda S, Rossi F (2017) Comparison among computed tomographic features of adipose massed in dogs and cats. Cet Radiol Ultrasound 58(1):29–37. https://doi.org/10.1111/vru.12445
Strafuss AC (1987) Necropsy: procedures and basic diagnostic methods for practicing veterinarians. Thomas C.C., Springfield, IL
Straw RC, Le Couteur RA et al (1989) Multilobular osteochondrosarcoma of the canine skull: 16 cases (1978–1988). J Am Vet Med Assoc 195:1764–1769
Straw RC, Powers BE et al (1996) Canine mandibular osteosarcoma: 51 case (1980–1992). J Am Anim Hosp Assoc 32:257–262
Sullivan DJ (1960) Cartilaginous tumors (chondroma and chondrosarcoma) in animals. Am J Vet Res 21:531–535
Sylvestre AM, Brash ML et al (1992) A case series of 25 dogs with chondrosarcoma. Vet Comp Orthop Traumatol 4:13–17
Szymendera J (1970) Bone mineral metabolism in cancer, 1th edn. Springer-Verlag, Berlin

Takai Y, Sano T, Watanabe T (2020) A case of hepatic leiomyosarcoma with osteosarcomatous differentiation (malignant mesenchymoma) in a dog. J Toxicol Pathol 33(1):33–37. https://doi.org/10.1293/tox.2019-0065

Tan ML, Choong PFM, Dass CR (2009) Osteosarcoma: conventional treatment vs. gene therapy. Cancer Biol Ther 8(2):106–117

Tas O, De Cock H, Lemmens P, Pool RR (2013) Synovial osteochondromatosis and sclerosing osteosarcoma in a cat. Vet Comp Orthop Traumatol 26(2):160–164

Thomas WB, Daniel GB et al (1997) Parosteal osteosarcoma of the cervical vertebra in a dog. Vet Radiol Ultrasound 38:120–123

Thompson KG, Pool RR (2002) Tumors of bones. In: Meuten J (ed) Tumors in domestic animals, 4th edn. Iowa State Press, Blackwell (IA), pp 245–317

Thompson E, Fauber AE, Pool RR (2016) Diagnosis and outcome of periosteal chondrosarcoma in two cats. J Am Anim Hosp Assoc 52(5):312–318

Thornburg LP (1979) Giant cell tumor of bone in a cat. Vet Pathol 16:255–257

Thrall DE (1981) Orthovoltage radiotherapy of oral fibrosarcomas in dogs. J Am Vet Med Assoc 179:159–162

Thurman GB, Mays CW et al (1973) Skeletal location of radiation-induced and naturally occurring osteosarcomas in man and dog. Cancer Res 33:1604–1607

Tien PC, Quan M, Kuang S (2021) Sustained activation of notch signaling maintains tumor-initiating cells in a murine model of liposarcoma. Cancer Lett 494:27–39. https://doi.org/10.1016/j.canlet.2020.08.029

Tomita N, Sunden Y, Okamoto Y, Morita T (2017) Well-differentiated liposarcoma with chondroid metaplasia in the auricle of a dog. J Vet Med Sci 79(7):1236–1239. https://doi.org/10.1292/jvms.17-0168

Trigo FJ, Leathers CW et al (1983) A comparison of canine giant cell tumor and giant cell reparative granuloma of bone. Vet Pathol 20:215–222

Vail DM, MacEwen EG (2000) Spontaneously occurring tumors of companion animals as model for human cancer. Cancer Investig 18:781–792

Vanel D, Henry AM, Lumbroso J, Lemalet E, Couanet D, Piekarski JD et al (1984) Pulmonary evaluation of patients with osteosarcoma: role of standard radiography, tomography, CT, scintigraphy and tomoscintigraphy. Am J Roentgenol 143:519–523

Vanel M, Blond L et al (2012) Imaging of primary bone tumors in veterinary medicine: which differences? Eur J Radiol 82:2129–2139

Vascellari M, Tasca S et al (2007) Vertebral polyostotic lymphoma in a young dog. J Vet Diagn Investig 19:205–208

Vinayak A, Worley DR, Withrow SJ, Adams DS, Powers BE (2018) Dedifferentiated chondrosarcoma in the dog and cat: a case series and review of the literature. J Am Anim Hosp Assoc 54(1):50–59. https://doi.org/10.5326/JAAHA-MS-6566

Viswanathan S, Jambhekar NA (2009) Metastatic Giant cell tumor of bone. Clin Orthop Relat Res 468(3):827–833

Vlychou M, Athanasou NA (2008) Radiological and pathological diagnosis of paediatric bone tumours and tumour-like lesions. Pathology (Phila) 40:196–216

Waltman SS, Seguin B et al (2007) Clinical outcome of non-nasal chondrosarcoma in dogs: thirty one cases (1986–2003). Vet Surg 36:266–271

Wang FI, Liang SL et al (2005) Disseminated liposarcoma in a dog. J Vet Diagn Investig 17:291–294

Watine S, Hamaide A et al (2003) Resolution of chylothorax after resection of rib chondroma in a dog. J Small Anim Pract 44:546–549

Webb JA, Liptak JM, Hewitt SA, Vince AR (2009) Multilobular osteochondrosarcoma of the os penis in a dog. Can Vet J 50(1):81–84

Weber N, Tebeau C (1998) An unusual presentation of multiple myeloma in two cats. J Am Anim Hosp Assoc 34:477–483

Weishaar KM, Edmondson EF, Thamm DH, Olver CS (2014) Malignant mesenchymoma with widespread metastasis including bone marrow involvement in a dog. Vet Clin Pathol 43(3): 447–452. https://doi.org/10.1111/vcp.12180

Werner M (2006) Giant cell tumour of bone: morphological, biological and histogenetical aspects. Int Orthop 30:484–489

Wesselhoeft AL, Berg J et al (1991) Fibrosarcoma of the canine appendicular skeleton. J Am Anim Hosp Assoc 27:303–309

Wherle-Martinez AS, Dittmer KE, Aberdein D, Thompson KG (2016) Osteocalcin and Osteonectin expression in canine osteosarcoma. Vet Pathol 53(4):781–787. https://doi.org/10.1177/0300985815626574

White VA, Fanning CV, Ayala AG et al (1988) Osteosarcoma and the role of fine needle aspiration: a study of 51 cases. Cancer 62(6):1238–1246

Willis RA (1967) Pathology of tumors, 4th edn. Butterworths, London

Withrow SJ, Powers BE, Straw RC, Wilkins RM (1991) Comparative aspects of osteosarcoma. Dog *versus* man. Clin Orthop Relat Res 270:159–168

Wold LE, Adler CP, Sim FH, Unni KK (2003) Atlas of orthopaedic pathology, 2nd edn. Saunders, Philadelphia, pp 179–185

Wolke RE, Nielsen SW (1966) Site incidence of canine osteosarcoma. J Small Anim Pract 7:489–492

Wood BC, Grant DC et al (2002) What's your diagnosis? J Am Vet Med Assoc 221:939–940

Woog J, Albert DM et al (1983) Osteosarcoma in a phthisical feline eye. Vet Pathol 20:209–214

Wright IM, Minshall GJ (2012) Clinical, radiological and ultrasonographic features, treatment and outcome in 22 horses with caudal distal radial osteochondromata. Equine Vet J 44:319–324

Wrigley RH (2000) Malignant versus nonmalignant bone disease. Vet Clin North Am Small Anim Pract 30:315–347

Wülling M, Engels C et al (2001) The nature of giant cell tumor of bone. J Cancer Res Clin Oncol 127:467–474

Wülling M, Delling G et al (2003) The origin of the neoplastic stromal cell in giant cell tumor of bone. Hum Pathol 34:983–993

Yamamoto S, Yamamoto Y et al (2005) Extraskeletal myxoid chondrosarcoma in a cow. Vet Rec 156:842–843

Yarmish G, Klein MJ et al (2010) Imaging characteristics of primary osteosarcoma: nonconventional subtypes. Radiographics 30:1653–1672

Yildiz F, Gurel A et al (2003) Frontal chondrosarcoma in a cat. J Vet Sci 4:193–194

Zachary JF, Mesfin MG et al (1981) Multicentric osseous hemangiosarcoma in a Chianina-Angus steer. Vet Pathol 18:266–270

Zaruby JF, Williams JW et al (1993) Periosteal osteosarcoma of the scapula in a horse. Can Vet J 34:742–744

Zheng MH, Robbins P, Xu J, Huang L, Wood DJ, Papadimitriou JM (2001) The histogenesis of giant cell tumour of bone: a model of interaction between neoplastic cells and osteoclasts. Histol Histopathol 16:297–307

Zimmerman K, Almy F, Saunders G, Crisman M, Leonardi L (2019) An unusual case of giant cell tumor of soft parts in an American warmblood horse. Open Vet J 9(1):44–48

Chapter 3
Tumor-like Lesions of the Musculoskeletal System

3.1 Fibrous Dysplasia

Fibrous dysplasia is a tumor-like bone lesion in the past also referred as osteitis fibrosa and characterized as a malformation of the bone tissue that can involve the bone in monostotic or polyostotic. These forms frequently involving the bone marrow with the presence of fibrous connective tissue and woven bone, where typical lining osteoblasts lack, as like as the maturation in lamellar bone. This condition expects differential diagnosis with ossifying fibroma or, in other way, with osteoma and with bony cysts and is described mostly in young animals of different species: horses, dog, feline (enclosed domestic cats and Siberian tiger) (Fig. 3.1).

Clinical features referring a painful bone expansile lesion, often associated to pathologic fracture. Frequently may remain also asymptomatic especially when arise in the monostotic cases. In polyostotic forms the lesions are also typically described as "Cafè au lait spoits" as referred in human species also by Mario Campanacci and where the lesions vary in shape and size from small to destructive and diffuse with irregular borders also called "Coast of Maine." The bone fibrous lesion may be also responsible of secondary complications such as obstruction, compression, of adjacent tissues and organs. In animals it is also described in the craniofacial bones, in ribs, long bones, etc. Radiographs usually show a localized non-aggressive lesion with well-defined clear margins and frequent sclerotic peripheral rim. Sometimes is possible to detect also the presence of cartilaginous tissue that can cause the deformity of involved bone and that also may be diagnostic when present. **Gross features** of the lesion reveal a malformation of bone characterized by substitution of normal tissue with a compact-dense fibrous tissue mixed with bone, focally also mineralized and occasionally containing cartilaginous lobules.

Histological features of fibrous dysplasia are represented by varying proportions of osseous and fibrous tissues not bordered by osteoblasts. Osseous tissue is composed by a mixture of components represented by irregular trabeculae of woven

L. Leonardi, *Bone Tumors in Domestic Animals*,
https://doi.org/10.1007/978-3-030-90210-0_3

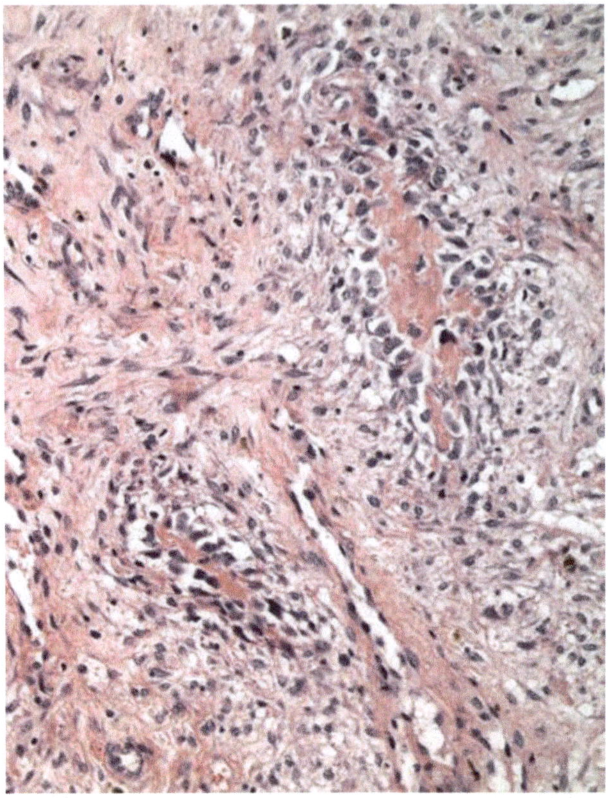

Fig. 3.1 Histopathology image of a tissue section from an ossifying fibroma that was replacing the rostral mandibular cortex in a yearling horse. In this lesion the bony trabeculae are lined by osteoblasts immediately as they arise from this fibrous tissue mass. (Courtesy of Dr. Roy R. Pool, DVM, PhD Emeritus Professor of Pathology UCVD and Texas A&M—USA)

bone, where are detected spindle-shaped cells and a rarely low number of osteoblasts. The lesion can also contain nodules of cartilage (usually hyaline) with phenomena of endochondral ossification. It is also possible to detect the presence of multinucleated osteoclast-like giant cells (probably due to increased production of IL-& by modified connectival cells), foam cells, and mineralized osteoid in a psammomatous arrangement that can orient the wrong diagnosis versus a case of meningioma.

It has now been widely demonstrated that fibrous dysplasia is associated with a mutation in the GNAS gene. **Immunohistochemistry** does not express specific immunophenotype also if spindle components result always positive for vimentin (Fig. 3.2).

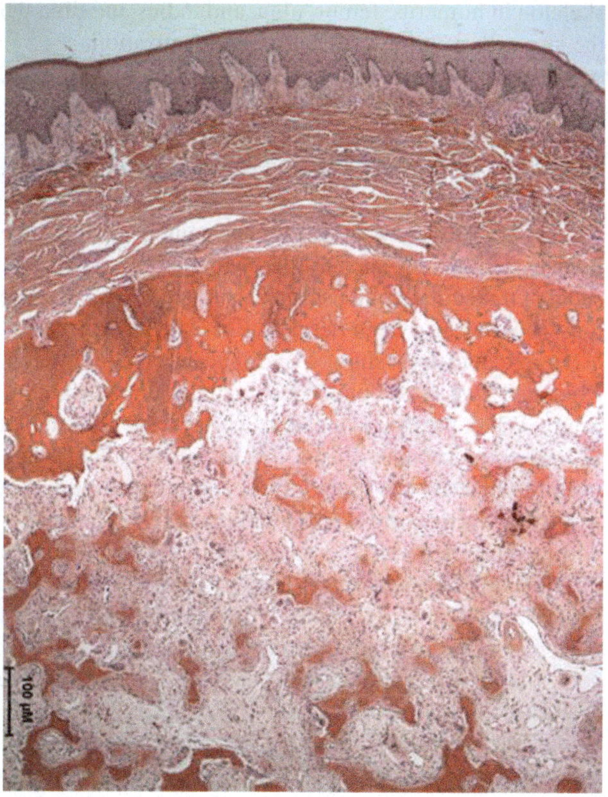

Fig. 3.2 Fibrous dysplasia creates a deforming jaw lesion in a 6-month-old dog in which osteoclasts remove the endosteal cortical bone as an expansile fibrous mass that replaces the cortical bone. The fibrous mass contains bony spicules that are not initially lined by osteoblasts as they arise in the fibrous tissue mass (Courtesy of Dr. Roy R. Pool, DVM, PhD Emeritus Professor of Pathology UCVD and Texas A&M—USA)

3.2 Solitary Bone Cyst

Solitary bone cyst, also called unilocular bone cyst or simple bone cyst, is an expanding osteolytic benign cavity fluid-filled cyst, usually localized in the wall of the involved bone and lined by a fibrous membrane with different thicknesses. The fluid contained in the cyst is normally clear and serous or rarely sanguineous. The etiology of solitary bone cyst is still unknown and occurs mostly in the long bones of young dogs. Etiopathogenetic hypotheses have considered that an obstruction of a vessel that causes a reduction in venous outflow may be responsible for an increase in the relative intraosseous pressure, with stimulation of the production of various cytokines that in turn stimulate bone resorption and the consequent formation of bone cysts. This could also explain the significant and constant presence of numerous osteoclastic cells always associated with this type of injury.

In 2016, Berger et al. report also a case of malignant transformation of a unicameral bone cyst of the humerus/shoulder in a 3-year-old Norwegian forest cat. Solitary bone cysts were also described in horses and also in young childrens (bone in rapid growth) where males are more involved than females, as like as appear in dogs, especially large breeds. Most involved bones in dogs are in appendicular skeleton in humerus, femur, ulna, and radius and tibia where metaphysis/diaphysis is the site with more frequent primary localization.

Clinical features: often clinical data referring of an asymptomatic disease (in most cases regress spontaneously) and only rarely painful, sometimes associated with pathologic fracture. Usually solitary bone cyst undergoes surgical curettage with filling of the cavity with a bone grafting (spongiosa/cancellous bone). A rarer variant is also described as aneurysmal bone cyst, but certainly important from a clinical point of view which is represented by bone aneurysmal cysts in which the bone cystic cavities appear filled with hemorrhagic fluid and often also separated by septa of trabecular bone tissue. This form is particularly common in Doberman breed dogs, but has also been described in cats and horses.

Image features show a radiolucent, well-demarcated metaphyseal/diaphyseal expanded cyst with bone cortex usually thinned and also with frequent medullary origin. When the cyst is localized close to the growth plate is also called "active" and instead when separated from this region is called "latent or inactive." MRI may be useful to identify the bloody fluid content or fluid nature of the cyst.

Gross features are representative of well delineated unicameral bone cavity filled with a fluid of a variable nature from serous to serous-hemorrhagic and lined by different thicknesses of connective fibrous tissue. When the cyst is localized in central area of marrow cavity with smooth wall the cortex is usually intact. On the other hand, the cyst may be also separated and subdivided by bone trabecules that contribute to giving the lesion a typical aspect also called "fallen leaf."

Histological features of the cystic lesion are represented by a wall containing connective fibrous tissue with fibroblasts mixed with fibrin and collagen fibers. Internal septae are constituted mostly by connective tissue too mixed with reactive new formed woven bone and scattered multinucleated osteoclastic giant cells, hemosiderin. Frequently is possible to detect the presence of foamy histiocytes and cholesterols deposits. When pathologic fracture occurs a prominent callus may be detected.

Differential Diagnosis and Prognostic Factors Certainly, in differential diagnosis it is necessary to contemplate some forms of bone lesions potentially similar to single bone cysts such as aneurysmal cysts and fibrous dysplasia. In the case of aneurysmal bone cysts, is possible that a single bone cyst can be confused with an aneurysmal one, especially if it has caused a pathological fracture. Bone aneurysmal cysts are lytic but more expandable than single ones. In single bone cyst the cortex is never expanded eccentrically as instead occur in case of aneurysmal bone cyst. In case of presence of material similar to cementum it may give an appearance similar to fibrous dysplasia lesions.

Prognostic factors reported in humans and animals describe a low-grade recurrence occurring in young subjects with large size of bone cysts. There are few reports of spontaneous healing of pathological fracture due to bone cyst but is also report a description of malignant transformation of a unicameral bone cyst in a cat (Figs. 3.3 and 3.4).

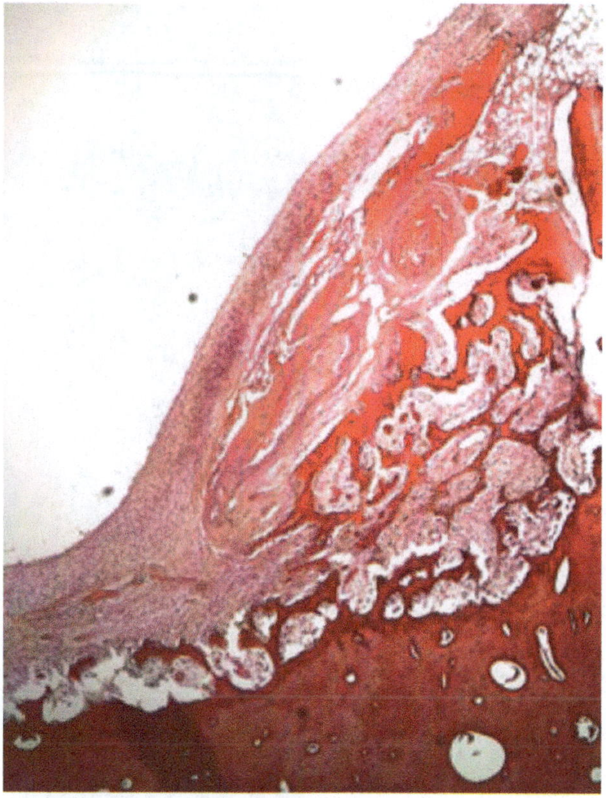

Fig. 3.3 Low magnitude histopatholgy image lining of the wall of this same unicameral bone cyst (Courtesy of Dr. Roy R. Pool, DVM, PhD Emeritus Professor of Pathology UCVD and Texas A&M—USA)

3.3 Juxtacortical Bone Cyst

Juxtacortical bone cyst is a typical lesion of the phalanx reported in horses localized near the joint and in the subchondral bone and characterized by a cystic lesion with fibrous and myxoid changes. In human species juxtacortical bone cysts are also reported in association with periosteal chondroid tumor of the cuboid and secondary to hyperparathyroidism.

3.4 Epidermoid Cyst and Epidermoid Cyst of the Phalanx

Epidermoid cyst, also called intraosseous epidermal cyst or epidermoid inclusion cyst, is a rare solitary benign bone cystic lesion lined by keratinized squamous multilayer epithelium described in dogs, horses, and humans. These lesions are usually found in acral skeleton and in particular in the phalanx or less often in the skull bone especially in human species. **Clinically** epidermoid cyst is frequently asymptomatic and has characteristics of swelling and pain, rarely accompanied with

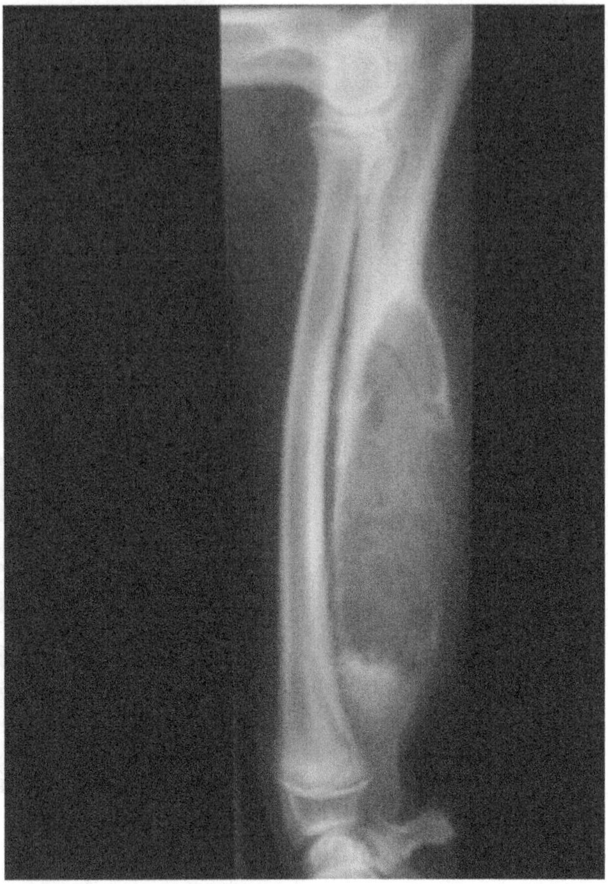

Fig. 3.4 Radiograph of a solitary bone cyst in the ulna of a young mature dog (Courtesy of Dr. Roy R. Pool, DVM, PhD Emeritus Professor of Pathology UCVD and Texas A&M—USA)

pathological fracture. The simple curettage is the main suggested treatment with very good prognosis. **Image findings** with radiographs revealed a well-defined lytic mass frequently with a bubble pattern due to fast expansion and with sclerotic margins. **Gross patterns** detect a mass containing caseous keratinous material with thinning of bone cortex and without periosteal reactivity.

Histopathology shows a cystic lesion where the cavity of the cyst wall results lined with squamous benign keratinized epithelium. Inside the cystic cavity results filled by debris of keratin as like as appear in human epidermal inclusion cysts. WHO suggesting to differentiate epidermoid cyst of the phalanx with squamous cell carcinoma but there are also other differential diagnosis as like as a reparative giant cell granuloma that may involve small bones but that has a very clear and easy histological features, and also some forms of osteomyelitis and also glomus tumors that maybe very similar to epidermoid cyst but only radiographically (Fig. 3.5).

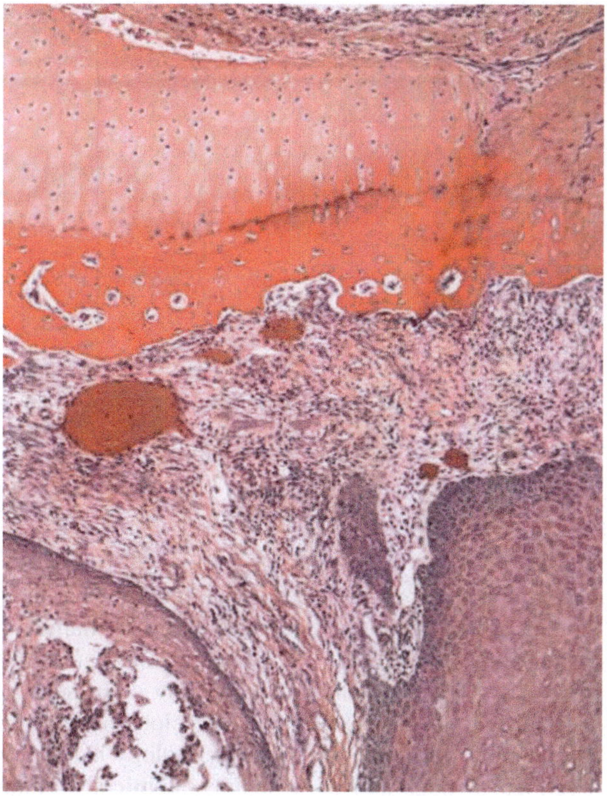

Fig. 3.5 Histopathology of an epidermoid cyst in P3 undermining the subchondral bone of the P2-P3 joint of a dog (Courtesy of Dr. Roy R. Pool, DVM, PhD Emeritus Professor of Pathology UCVD and Texas A&M—USA)

3.5 Myositis Ossificans

Myositis ossificans is a benign reactive modification of the parosteal soft tissues of the bone where the lesion results in a process of metaplasia and proliferation of bone and cartilage. This is not very appropriate terms to define this disease for a number of factors. In myositis ossificans inflammation is very rare, the muscle tissue must be involved, and especially in the first stage of the disease there is not deposition of new bone. With CT examination the localized lesion acquires a typical conformation called triple layers characterized by the presence of internal proliferating connective tissue, of osteoid and fibrous in the intermediate zone and mature bone in the external part of the involved tissue. Also, cartilaginous tissue may also be present. The disease was described in dogs, cats, bats, and pigs (Fig. 3.6).

Myositis ossificans recognizes a pathogenesis frequently due to traumatic injuries but there also cases of idiopathic myositis ossificans of unknown origin. Differential diagnosis considers histologically cases of extraosseus osteosarcoma and/or pseudomalignant bone tumors of the soft tissues, but with good experience may be easy to make the correct diagnosis of localized myositis ossificans.

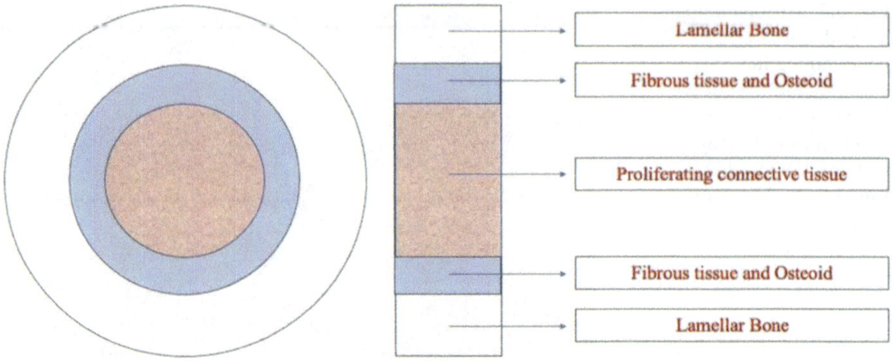

Fig. 3.6 The three layers of myositis ossificans

3.6 Villonodular Synovitis

Villonodular synovitis also called pigmented villonodular synovitis is a particular morbid process characterized by a diffuse inflammatory nodular brown colored (hemosiderin granules) proliferation of joint and tendons with particular involvement of synovial epithelium and neighboring mesenchymal tissues.

This kind of disease can cause erosions of bone from secondary proliferative involvement from the joint capsule. Etiology is still unknown and the disease was described in dogs and goats mainly. **Clinically** villonodular synovitis manifests swelling, pain, and restriction in movement due to the affected joint with increase of the synovial fluid and frequently appearance of periarticular erosions and sclerosis.

Gross features of the disease are represented by brownish villous to nodular thickened of synovial membrane. **Histology:** histologically is possible detected a package of villous lesions of the synovia and a mixture of macrophages and lymphocytes. The numerous villonodular lesions result covered by synovial epithelium detectable in the outer layer of stromal internal connective tissue rich of blood-filled vessels in the wall. The lesions are usually high cellular with lymphocytes, macrophages (with brown pigment inside) with nuclei of different sizes and hyperchromatic, and plasma cells rarely accompanied also by multinucleated giant cells (diagnostic feature) (Fig. 3.7).

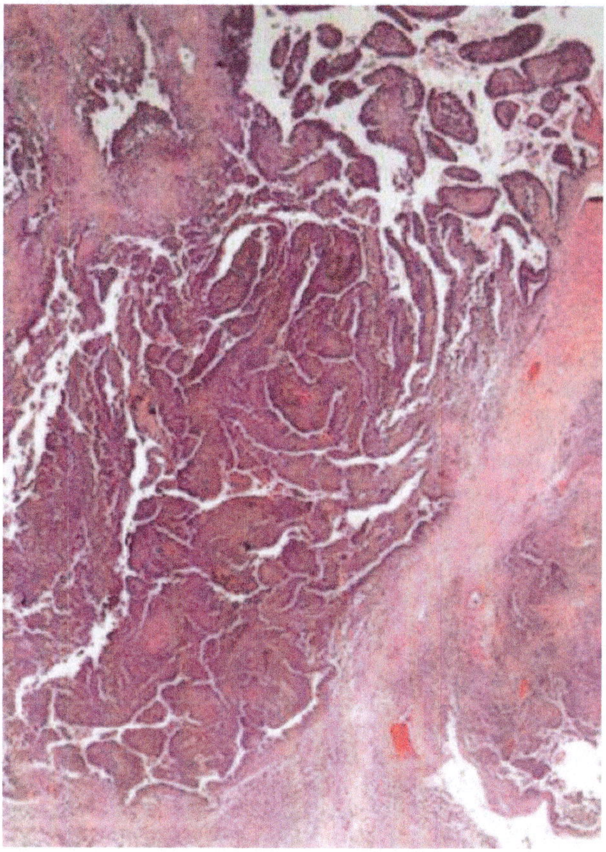

Fig. 3.7 Villonodular synovial lesion in the stifle joint of a dog. This is a copy of a poor illustration that I received on my computer screen but it shows the nodularity of villi (Courtesy of Dr. Roy R. Pool, DVM, PhD, Emeritus Professor of Pathology UCVD and Texas A&M—USA)

Bibliography

Akerblom S, Sjöström L (2006) Villonodular synovitis in the dog: a report of four cases. Vet Comp Orthop Traumatol 19(2):87–92

Carrig CB, Seawright AA (1969) A familial canine polyostotic fibrous dysplasia with subperiosteal cortical defects. J Small Anim Pract 10:397–405

Chapurlat RD, Meunier PJ (2000) Fibrous dysplasia of bone. Baillieres Best Pract Res Clin Rheumatol 14:385–398

Feldman F (1995) Tuberous sclerosis, neurofibromatosis, and fibrous dysplasia. In: Resnick D (ed) Diagnosis of bone and joint disorders, vol 6, 3rd edn. W.B. Saunders, Philadelphia, pp 4353–4395

Feller L, Wood NH et al (2009) The nature of fibrous dysplasia. Head Face Med 5:22

Frank H, Kirchhof N, Biesenbach W, Frese K (1995) Intraosseous epidermoid cysts of the toe phalanx in dogs. Berl Munch Tierarztl Wochenschr 108:73–76

Fraser B, Else RW, Jones E (2006) Intraosseous epidermoid cyst of the third phalanx in a thoroughbred gelding. Vet Rec 159:360–362

García-González M, Muñoz Guzón FM, González-Cantalapiedra A, López-Peña M, de Frutos PF, Pereira-Espinel Plata T, González Fernández PM, Serra Rodríguez JA (2021) Case report: first

evidence of a benign bone cyst in an adult Teckel dog treated with shark teeth-derived Bioapatites. Front Vet Sci 8:626992. https://doi.org/10.3389/fvets.2021.626992
Guilliard MJ (2001) Fibrodysplasia ossificans in a German shepherd dog. J Small Anim Pract 42(11):550–553. https://doi.org/10.1111/j.1748-5827.2001.tb06026.x
Hausmann JC, Manasse J, Churgin S, Steinberg H, Clyde VL, Wallace R (2016) Myositis ossificans traumatica in a vampire bat (Desmodus rotundus). J Zoo Wildl Med 47(3):895–899. https://doi.org/10.1638/2015-0289.1
Headley SA, Kummala E, Saarinen H, Tupamäki A, Tulamo RM (2009) Diagnostic exercise: intraosseous epidermoid cysts in the third phalanx of a dressage mare. Vet Pathol 46(2):355–357
Homer BL, Ackerman N et al (1992) Intraosseous epidermoid cysts in the distal phalanx of two dogs. Vet Radiol Ultrasound 33:133–137
Jones NY, Patterson-Kane JC (2004) Fibrous dysplasia in the accessory carpal bone of a horse. Equine Vet J 36:93–95
Liu SK, Dorfman HD (1974) Intraosseous epidermoid cysts in two dogs. Vet Pathol 11:230–234
Mapuvire T, Kandiwa E, Mbiri P, Samkange A, Madzingira O, Mushonga B (2020) Chronic lymphoplasmacytic villonodular proliferative synovitis in a 10-year-old Jack Russell terrier dog. Int J Vet Sci Med 8(1):100–105. https://doi.org/10.1080/23144599.2020.1842038
Marti JM (1997) Bilateral pigmented villonodular synovitis in a dog. J Small Anim Pract 38(6): 15–17. https://doi.org/10.1111/j.1740-8261.1998.tb00319.x
Milgram JW (1990a) Nonsystemic dysplasias of cartilage growth. In: Milgram JW (ed) Radiologic and histologic pathology of nontumorous diseases of bones and joints, vol 1. Lea and Febiger, Philadelphia, pp 113–130
Milgram JW (1990b) Polyostotic fibrous dysplasia. In: Milgram JW (ed) Radiologic and histologic pathology of nontumorous diseases of bones and joints, vol 1. Lea and Febiger, Philadelphia, pp 201–213
Morton BA, Hettlich BF, Pool RR (2015) Surgical treatment of traumatic myositis Ossificans of the extensor carpi Radialis muscle in a dog. Vet Surg 44(5):576–580. https://doi.org/10.1111/j.1532-950X.2014.12297.x
Ravanetti P, Lechartier A, Hamon M, Zucca E (2021) A composite absorbable implant used to treat subchondral bone cysts in 38 horses. Equine Vet J. https://doi.org/10.1111/evj.13428
Riminucci M, Robey PG et al (2007) The pathology of fibrous dysplasia and the McCune-Albright syndrome. Pediatr Endocrinol Rev 4:401–411
Roode SC, Shive HR, Hoorntje W, Bernard J, Stowe DM, Pool RR, Grindem CB (2018) Multiloculated solitary (unicameral) bone cyst in a young dog. Vet Clin Pathol 47(3): 484–488. https://doi.org/10.1111/vcp.12618
Soltero-Rivera M, Engiles JB, Reiter AM, Reetz J, Lewis JR, Sánchez MD (2015) Benign and malignant proliferative fibro-osseous and osseous lesions of the oral cavity of dogs. Vet Pathol 52(5):894–902. https://doi.org/10.1177/0300985815583096
Vagias M, Cassidy JP, Skelly C, Mullins RA (2020) Intraosseous epidermoid cysts of adjacent digits in a dog. BMC Vet Res 16(1):1–9. https://doi.org/10.1186/s12917-020-02545-7
Wilson RB (1989) Monostotic fibrous dysplasia in a dog. Vet Pathol 26:449–450
Winbladh K, Fransson BA, Svensson G, Karlstam E, Uhlhorn M (2020) Aneurysmal bone cyst in the pelvis of a cat: successful outcome of partial ileectomy with limb preservation. JFMS Open Rep 6(2):2055116920974984. https://doi.org/10.1177/2055116920974984

Chapter 4
General Considerations About Bone Tumors

One of the first medical diagnostic findings in the laboratory in the condition of a suspected bone tumor lesion can be represented by the highlighting of a state of hypercalcemia, a dangerous condition for the heart (cardiotoxic activity) and for its potential cause of consequent dystrophic calcifications. In all cases in which hypercalcemia is detected, it is always recommended to perform radiographic investigations to evaluate osteolytic lesions of the skeletal system. In the absence of specific bone lesions, is necessary to evaluate other hypercalcemic diseases, not necessarily of bone, both metabolic and toxicological, especially those in close correlation with chlorine and potassium homeostasis and parathyroid function. In case of hypercalcemia in dogs and cats it will be necessary to consider the following diseases in differential diagnoses: bone tumors or inflammatory-degenerative bone diseases, primary hyperparathyroidism and pseudo-hyperparathyroidism, severe dehydration, renal insufficiency, and hypoadrenocorticism.

Bone tumors as well as some forms of osteitis and periosteitis can cause significant changes in bone density that can lead to frequently localized pictures, consisting essentially of focal osteopenia.

As regards the clinical staging of malignant bone tumors, being very rare in them a process of metastasis involving the pathway of the lymphatic circulation with involvement of the lymph node structures, it will be strictly limited to the detailed clinical and morphological description of the primary bone tumor (Tables 4.1 and 4.2) and its distant metastases through the evaluation system called TM:

PRIMITIVE TUMOR—T index.
$\mathbf{T_0}$—no evidence of bone tumor.
$\mathbf{T_1}$—bone tumor localized to medullary canal or cortex.
$\mathbf{T_2}$—bone tumor which expands beyond the periosteum.

Multiple bone tumors and/or polyostotic forms must be evaluated and described separately.

L. Leonardi, *Bone Tumors in Domestic Animals*,
https://doi.org/10.1007/978-3-030-90210-0_4

Table 4.1 Surgical stages for human **benign** musculoskeletal tumors (Enneking 1983, modified)

Stage	Grade	Site	Metastases	Definition	Behavior
1	G_0	T_0	M_0	Latent or inactive	Remains static or heals spontaneously
2	G_0	T_0	M_0	Active	Progressive growth but limited by natural barriers
3	G_0	T_0	M_0	Aggressive	Progressive growth, not limited by natural barriers

Table 4.2 Surgical stages for human **malignant** musculoskeletal tumors (Enneking 1983)

Stage	Grade	Site	Metastases
I			
A	Low (G_1)	Intracompartmental (T_1)	None (M_0)
B	Low (G_1)	Extracompartmental (T_2)	None (M_0)
II			
A	High (G_2)	Intracompartmental (T_1)	None (M_0)
B	High (G_2)	Extracompartmental (T_2)	None (M_0)
III			
A	Low (G_1)	Intra or extra compartmental (T_1–T_2)	Regional or distant (M_1)
B	High (G_2)	Intra or extra compartmental (T_1–T_2)	Regional or distant (M_1)

DISTANT METASTASIS—M index.
M_0—no evidence of distant metastasis.
M_1—presence of distant metastasis with detailed and specific single or multiple localization.

HISTOLOGICAL GRADE—G index.
G_1—low-grade tumor.
G_2—high-grade tumor.

STAGING—stage index.
Stage I—G_1M_0.
Stage II—G_2M_0.
Stage III—$G_{1/2}M_1$.

Substage a—T_1.
Substage b – T_2.

I would like to report also what established in human medicine by Enneking et al. in 1983, reported also by Mario Campanacci and related for benign and malignant musculoskeletal tumors surgical stages as a potential template for future classifications of bone tumors of domestic animals (Tables 4.1 and 4.2).

(a) Core biopsy and fine-needle aspiration are the frequent methods used to obtain samples for pathological diagnosis. The tissue sample taken by this procedure is frequently limited resulting in difficulty for osteosarcoma diagnosis. OSA is a great histologic mimicker and poses the diagnostic challenge especially in small tissue biopsies. Histopathologically and cytologically, the diagnosis of OSA relies on the presence of malignant tumour cells and neoplastic bone. Tumour cells in OSA typically demonstrate severe anaplasia and pleomorphism and may be epitheloid, plasmacytoid, fusiform, small, and round or spindled. The neoplastic bone is eosinophilic when unmineralized (osteoid) and basophilic/purple if mineralized. The osteoid matrix appears solid, homogeneous, amorphous, irregular, and lace-like or curvilinear in the background or between tumour cells. In some cases, distinguishing unmineralized matrix (osteoid) from other eosinophilic extracellular materials especially collagen may be difficult and subjective. In order to make an accurate diagnosis, it is important to find molecular marker to distinguish OSA and other osseous primary tumors.

Osteocalcin (OCN) is known to be a bone tissue-specific protein and in several cases may be helpful to differentiate osteosarcomas from other tumors especially from other not producing mineral matrix.

Alkaline phosphatase (ALP) is an enzyme present in all tissues particularly concentrated in the liver, kidney, placenta, and bone. In the musculoskeletal apparatus, ALP is abundant in osteoblasts cells with a primary role in the mineralization of newly formed bone. In some researches conducted on canine, ALP is a sensitive and specific useful marker to identifying tumor cells in OSA.

4.1 Specimen Selection, Processing, and Histologic Methods for Evaluating Bone Specimens

This part summarizes basic and necessary procedures for the selection and processing of bone tissue specimens. The reader is referred to and may wish to consult with books of histological techniques available in the specific veterinary medical literature.

Unlike the common pathology and histopathology laboratories, the structures for the processing and treatment of bone tissue specimens must necessarily have special laboratory facilities and special equipment that can allow the best results from hard tissue processing. The facilities and equipment to operate a bone tissue processing laboratory include: are a surgical dissection bench or table, bandsaw, and rotary saw for cutting thin bone sections for radiologic exam and decalcification prior to tissue site selection for histopathologic examination, access to a radiographic instrument (s) for preparing gross images, and high detail radiographic images of thin 2–4 mm thick bone sections.

The basic surgical instruments and supplies include: scalpel, scissors, forceps and also by gloves, surgical towels, plastic bags, etc., with also saws (handsaw, circular saw, diamond wire saw, etc.) and grinding systems.

Basic procedures for bone tissues: include different processes such as fixation, decalcification, processing of dehydration and embedding, sectioning, and staining.

It is advisable to carry out the fixation process (preventing tissue decomposition with preservation of the cells) as quickly as possible after removing the sample from the organ. The choice of fixative depends on the type of tissues and the type of histopathological investigation that will be carried out. The fixative most used for histological investigations is represented by 10% neutral buffered formalin that is working very good also for bone and cartilage tissue where can penetrate rapidly and continuously. 10% neutral buffered formalin permits to well stain with the most common techniques. There are many other fixative available and less common of formalin and they are represented by Ethanol, Bouin's solution, Carnoy's fluid, Glutaraldehyde, alkaline phosphatase, and acid phosphatase. A good fixation is the most important key step in bone histopathological investigations. The size of the samples must be optimized and the suggested fixative/specimen ratio is >10:1. Specimens less of 5–6 mm can fix well in between 24–48, while specimens larger in size needs also 72 h of fixation or in some case perfusion fixation.

To remove water and embedded tissue is necessary to bathing in a series of alcohols (generally 50%, 70%, 80%, 95%, 100%, or absolute) but in most cases for bone tissues is required a preventive treatment of decalcification.

To obtain a good 3–5 μm section of bone tissue there are several procedures not always available in all laboratories. The paraffin used to embed tissues is softer than most bones and it must be pre-treated to make the bone tissues softer in order to be sectioned. Most common solution for decalcification includes Formic acid, Hydrochloric acid, and EDTA (Ethylene Diamine Tetra Acetic Acid).

The composition of bone is summarized in: Calcium (in crystals of inorganic salts) + bone matrix (organic), where hydroxyapatite $(Ca_{10}(PO4)_6(OH)_2$ represents the inorganic component that must be removed by decalcification that must reach the "end point" when salts has been removed completely. Remember that the higher risk to treat the bone tissue with strong acids and related mixtures is to over decalcify the bone. The choice of decalcifying fluids is various and also their related applications are strongly dependent to the bone tissue in exam.

Procedures to decalcify bone involve the immersion of samples in acid solutions or in acid-chelator complex solutions. The process should be facilitated by ultrasonic stimulation, vacuum or agitation to reduce the time of exposure and manipulation of bone tissues.

The solution used in our routinely methods in the Department of Veterinary Pathology (Veterinary Medicine) at the University of Perugia, is composed by a mixture of 50 parts (ml.) of Formic Acid (HCOOH 99% in aqueous solution), 40 parts (ml.) of hydrochloric acid (HCl 37% in aqueous solution), and 910 parts (ml.) of distilled water.

The common method to decalcify bone samples is immersion in the decalcifying or chelating solutions. The effects of decalcification with these compounds should be augmented by vacuum, heat, or ultrasonic, or simple agitation but always remember that the decalcification processes and its eventual systems to accelerate it must be kept under constant control to avoid over decalcification and loss of cellular detail.

4.1.1 *Determination of End Point Decalcification of Bone Using Three Techniques*

There are several methods to determine the end point of decalcification for bone or "mineralized" tissues after treatment using 3 modalities:

- **MANIPULATION:** when you have samples of a big size, the simplest and most used method is certainly the evaluation of the degree of folding of the decalcified bone tissue. A good bone decalcification treatment must be able to reduce the bone tissue to a consistency similar to that of rubber (Fig. 4.1). This method called also "probing and bending" is used to determining when the tissue is ready to be sectioned and the sample should be probing also with a needle or scalpel blade.
- **TIMED IMMERSION:** these methods should be used for standard size samples as with bone marrow biopsies where the constant and regular size of the samples can allow a kind of standardization of the decalcification's methods. An and Martin suggest that "a 5mm-thick slab of femoral head will be adequately decalcified to do full face sectioning after 24 hrs. in 250mL of a 1,35N hydrochloric acid/chelator solution."
- **RADIOGRAPHY:** according to some, this represents the best method for determining the total decalcification of the sample after the chemical treatment. However, to perform this method, is necessary to perform a preliminary X-ray of the starting bone tissue before treatment and then perform subsequent X-rays to monitor the softening process of the tissue. Obviously, this method, although valid and effective, requires longer evaluation times and relative higher

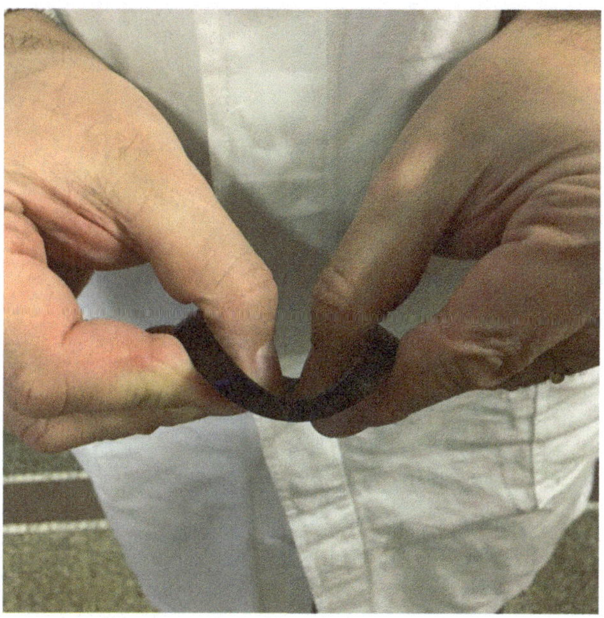

Fig. 4.1 Graphic example of the outcome of a bone decalcification process

management costs which also require the presence of special instruments in the laboratory.

Bibliography

Adler C-P (2000) Bone diseases: macroscopic, histological, and radiological diagnosis of structural changes in the skeleton, 1st edn. Springer, London

An YH, Martin KL (2010) Handbook of histology methods for bone and cartilage. Human Press, Totowa, NJ

Enneking WF (1983) Musculoskeletal tumor surgery. Churchill Livingstone, Edinburgh-London

Karcher SJ (2002) Laboratorio di Biologia Molecolare. Zanichelli Editori, Bologna

Sodikoff CH (1995) Medicina di Laboratorio del cane e del gatto. Masson Edizioni Veterinarie, Milano

Wold LE, Adler CP, Sim FH, Unni KK (2003) Atlas of orthopaedic pathology, 2nd edn. Saunders, Philadelphia, pp 179–185

Chapter 5
Small General Manual of Histological and Diagnostic Methodologies for Bone and Cartilage

In busy veterinary diagnostic service, a few of these tumors are observed in dogs but only given an accurate diagnosis. The cells of these tumors are an admixture of plump spindle and polyhedral cells that do not form matrix.

The enormous differences in the appearance and behavior of bone tumors sometimes make very difficult to reach a definitive and clear diagnosis that always need rather to have available a clinical, radiographic, and instrumental framework that allows to evaluate together with the histopathological examination also these aspects characterizing the bone lesions that, especially in case of histopathological evaluation of small biopsies where the diagnostic path sometimes could be misinterpreted.

5.1 Introduction

I like to introduce this chapter with the words of Dr. Albert C. Strafuss reported on his book of Necropsy, edited in 1987 always actual in the management of the lesions and necroscopic procedures: *However, one should never forget that the history and clinical examination are important factors in making a diagnosis. Laboratory results are of little or no value if the veterinarian (I add the pathologist) is unable to interpret the results and correlate them with the history and clinical picture. Laboratory tests should be used for diagnostic confirmation and not solely for diagnosis. Clinical laboratory findings should be utilized to support a necropsy as well as a clinical workup.* And more: *The quality of laboratory results are related to the quality of the samples collected and submitted.*

The manual skills and the diagnostic and experimental processing techniques applied to bone and cartilage tissues provide for a specific experience that goes beyond the classic routine and the now standardized methods of treatment of the so-called classic samples and tissues. The laboratory for the treatment of osteo-cartilage tissues also requires specific characteristics with tools that also allow the specific treatment of these tissues and the relative steps to treat them in the various

L. Leonardi, *Bone Tumors in Domestic Animals*,
https://doi.org/10.1007/978-3-030-90210-0_5

phases of treatment (fixation, decalcification, media infiltration, embedding, cut, etc.).

5.2 Diagnostic Process and General Investigation Protocols

It is now strongly supported by all osteopathologists that the issuance of the definitive diagnosis of many bone tumors is necessarily based on the comparative evaluation of clinical symptoms, the results of collateral investigations, including those performed through diagnostic imaging systems, and, above all, on the anatomo-histopathological evaluation of the samples.

The questionable practice, unfortunately still too widespread, of issuing a diagnosis only on the basis of radiographic findings (Fig. 5.1) can lead, with a high percentage of probability, to incurring serious diagnostic and interpretative errors of the lesion. In fact, it is not unlikely that the radiographic picture of an osteomyelitis

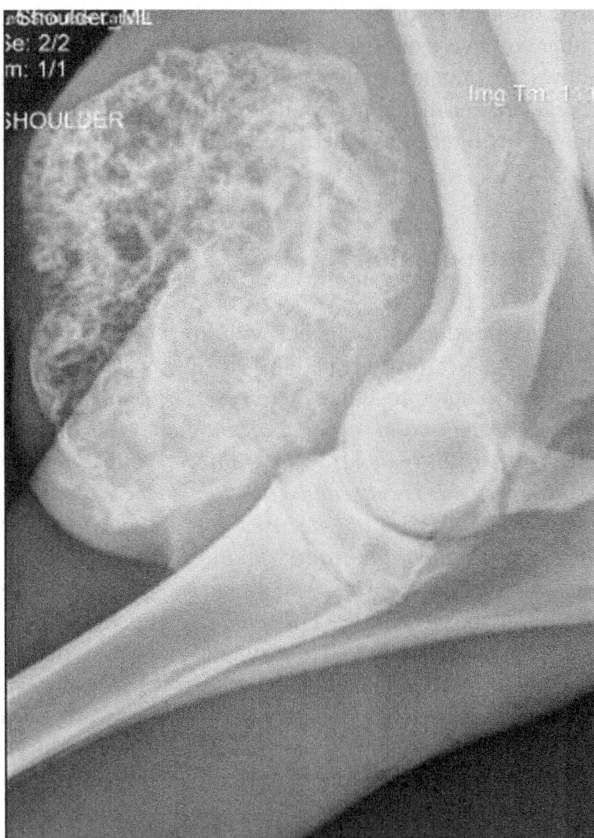

Fig. 5.1 Radiograph of lesions of ossifying myositis that arises in the fascia of the musculature of the shoulder of an immature horse and that interferes with movement (Courtesy of Dr. Roy R. Pool, DVM, PhD, Emeritus Professor of Pathology UCVD and Texas A&M—USA)

can be superimposed on that of an osteosarcoma, or vice versa, or that a bone cyst could be confused with a bone metastasis of an epithelial tumor, or vice versa.

5.3 Anamnestic Investigations and Clinical Symptoms

Species, sex, race, and age: the recording of these data is important for an initial epidemiological-probabilistic assessment of any neoplastic bone lesion. Primary tumors of the skeletal system, according to what is reported in the specialized literature, mainly affect the canine species and the feline species, with a slightly greater frequency in males, compared to females and larger breeds, of adult age, unlike the human species, where some bone tumors are more frequent in young subjects. For each neoplastic form, we will try, subsequently, to provide further epidemiological data.

Growth rate: generally, a rapid growth rate is the expression of a tumor with a high degree of malignancy, although there may be exceptions. In fact, in some non-neoplastic forms, such as some rare types of granulomas or bone aneurysmal cysts, growth can be as rapid as an osteosarcoma. In other rare soft tissue injuries, such as nodular fasciitis or ossifying myositis, the damage can arise and grow even faster than a malignant tumor. On the other hand, however, there are malignant neoplastic forms, such as synovial sarcoma, which can arise and remain in a latent-like state even for a long time (Figs. 5.1 and 5.2).

Pain: facilitates the diagnosis of a more likely malignant form and, therefore, infiltrating and destructive. For soft tissue tumors, pain is an almost constant feature of synovial sarcoma. The hyperthermia factor is not of much help in the anamnestic evaluation of the case as it may or may not be present in association with the tumor.

An accurate collection of data relating to the animal's history can help the pathologist to reconstruct the dynamism of the onset of the tumor and any predisposing factors of the neoplasm. The recording of the remote anamnesis involves the evaluation of all those data related to the history of the animal and its relatives, to any previous morbid diseases related to dysmetabolic and functional factors, as well as to any previous skeletal problems, such as trauma, infections, etc. The recent medical history, on the other hand, will tend to record all those factors that have clinically characterized the disease, the investigations performed, the diagnoses made, the assessment of predisposing factors (for example, obesity), the hematological and biochemical parameters, etc. It is also important not to neglect the recording of data relating to the environment where the animal lived, the type of diet, promiscuity with other animals, environmental treatments, etc.

Localization of the tumor: this evaluation is of fundamental importance as, for many bone tumors, the diagnosis can strictly depend also on the localization of the tumor. It is widely believed that, for example, periosteal tumors are biologically less malignant than their counterparts with intramedullary localization.

Radiology: the "morphological" radiographic evaluation of the lesion is an absolute essential part in the diagnosis of bone cancer. For this reason, the veterinary pathologist should take attention to diagnose a bone tumor if possible in first

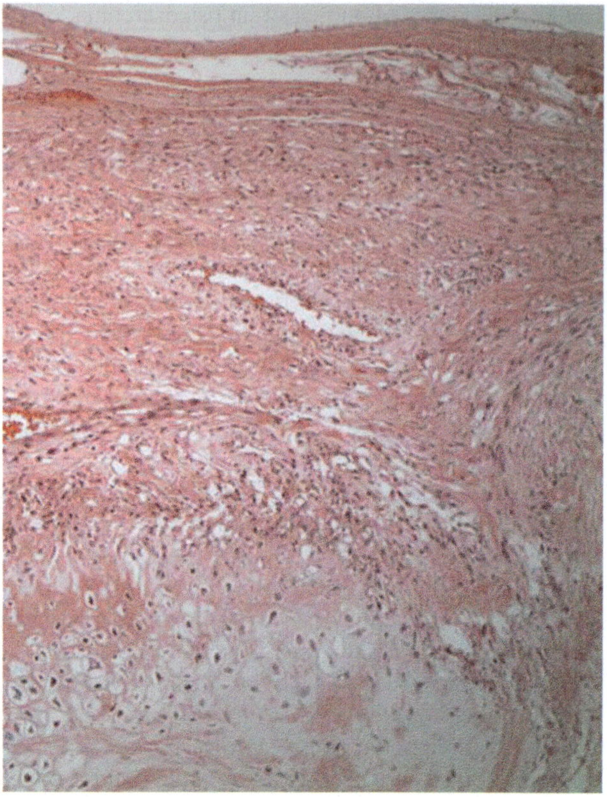

Fig. 5.2 Histologic section from part of a lesion of ossifying myositis forming an expanding mass located in intermuscular fascia and in this field contains a nodule of primitive cartilage embedded in proliferative fibrous tissue bordered by a fibrous capsule. Bony nodules were also present but not included in this figure (Courtesy of Dr. Roy R. Pool, DVM, PhD, Emeritus Professor of Pathology UCVD and Texas A&M—USA)

examination and evaluation of the radiographic aspects of the lesion. This appears even more important for malignant bone tumors, for which amputation-mutilation is often necessary. The histological section can sometimes be insufficient to ensure that a bone tumor is diagnosed with certainty. The importance of this concept is that, in order to reduce clinical diagnostic evaluation errors, in reaching the definitive diagnosis, it is always better to perform a collegial (clinical, radiologist, and pathologists) evaluation of the results of the performed evaluative investigations. The radiographs, as well as the images obtained by CT or MRI, should always be compared with the macroscopic aspects of the tumor and, finally, related to the subsequent histopathological findings of the lesion.

Radiographs provide a negative image of the tumor and reveal the degree of aggression and the type of response of the affected bone segment. There are two essential changes in the bone affected by a tumor: osteolysis and reactive osteogenesis.

An osteolytic area, with undefined edges and with cortex overtaken by tumor growth, and areas of poor reactive osteogenesis, indicates that the tumor has grown rapidly and is aggressive. An osteolysis with well-defined edges, with sclerotic margins and reactive osteogenesis, with preserved cortex or with a bony "shell" as a result of periosteal osteogenesis, indicates slow, non-infiltrating growth.

The X-ray image also represents a guide that allows you to better evaluate the path to follow for the execution of any biopsy sampling. In some cases, however, the radiographic image can even be more "diagnostic" than biopsies (parosteal osteosarcoma, peripheral chondrosarcoma, ossifying myositis, etc.), otherwise in soft tissue tumors, where radiographic images are of lesser importance. Nevertheless, also the presence of intratumoral calcifications can lead the diagnosis towards a synovial sarcoma.

Arteriography: has considerable value in defining the three-dimensional margins of the tumor and their relationship with the main vessels, also, if frequently this type of investigation has been replaced by CT or MRI. However, there are still important for a surgical-therapeutic evaluation, as they allow to evaluate the set of vessels within the neoplasm and also because they allow a hypothetical rough evaluation of the degree of malignancy of the tumor. In fact, the angiographic images of malignant tumors can show, with a significant frequency, characteristics of intense hypervascularity, anomalous, tortuous, ectatic, cavernous vessels. Indirect signs of arteriovenous shunts may also be present, while, nevertheless, in some malignant tumors, such as chondrosarcoma, necrotizing, hemorrhagic, and cystic tumors, areas without vessels may also be detectable.

Radionuclide Bone scan: this technique is performed with diphosphonate compounds labeled with Technetium 99 (TC99) which appears overexpressed (early phase) when the blood reserve of the bone is high and (late phase) when the salt crystals have formed, such as in a "area of osteogenesis." With this investigation, every process in which an active osteogenesis is present is accompanied by the formation of a "hot" area of positivity. The "bone scan" technique allows us to:

1. Evaluate an index of quiescence or activity of the tumor, even if not in an absolute sense, as some lesions could show "hot" areas in the course of restarting processes.
2. Examine the entire skeleton with a single investigation;
3. Reveal particular neoplastic localizations, not easily identifiable on radiographs;
4. Detect the extension of the tumor along the affected bone and its medullary canal;
5. Allows to evaluate skeletal reactions related to the presence of adjacent soft tissue tumors. This can allow you to decide how to take the samples, including or not including biopsies of the bone itself;
6. Allows to set up patient control protocols, after the therapeutic treatment;
7. Allows to obtain an important indication relating to the response of many osteosarcomas, resulting from the execution of preoperative chemotherapy protocols.

Computerized Axial Tomography (CT): this technique, known in the Anglo-Saxon language as CT (computerized tomography), allows you to obtain, through the use of X-rays, three-dimensional tomographic images of the affected bone segment. It is generally performed with the intravenous administration of a contrast medium, in order to also be able to visualize the blood vessels, the vascularization of the tumor and any reactive tissues. In bone oncology, it allows you to assess the extent of the tumor within the affected bone segment, the relationships that periosteal or parosteal tumors can establish with the underlying cortex; the extension of the tumor towards the soft tissues and the involvement of compartmental and extracompartmental areas; the relationships of continuity and contiguity with blood vessels and viscera; the involvement of joints, synovial capsule, joint spaces, etc. In soft tissue cancers, CT can reveal the location of the tumor in muscle compartments and extracompartmental spaces and its possible relationships with the vessels, bone, and viscera. It can also provide important information on the nature of the neoplastic mass: solid, liquid, or "adipose."

The tomography is better at MRI (magnetic resonance imaging) in the evaluation of mineralized tissue, such as bone and calcium deposits. This technique can be useful for the evaluation of the effects of radiotherapy and chemotherapy, for the precise and sensitive evaluation of lung metastases. However, it is not very useful when you have to evaluate a clear image of a bone in which there are metal elements, such as plates or prostheses. With this investigation technique, false positives can be represented by conditions such as: extension of the tumor within the medullary canal, which can be confused with edema and hemorrhages of the bone marrow or by lung metastases, which can be mistaken for nodules of atelectasis or fibrosis.

Magnetic Resonance Imaging (MRI): these images can provide sagittal, coronal, and axial planes of the bone under examination; in comparison with the CT provides better detail and tissues and a defined margin of muscle compartments; it does not involve the use of ionizing radiation and iodine contrasts and can be used several times, even in pregnant animals, without having adverse effects. In general, MRI resolution is superior to CT, for the study of soft tissues, bone marrow, and spinal column. The MRI with angiography, it can highlight the vessels both inside and around the tumor.

Ultrasound: can be useful in soft tissue injuries. It is particularly useful in differentiating cystic lesions from homogeneous solid masses. Otherwise, the images obtained with these investigations are not as accurate as those described for the CT and MRI.

Laboratory tests: these are important for the diagnosis of neoplastic bone diseases, such as multiple myeloma or plasmacytoma (monoclonal proteins), neuroblastoma (catecholamine metabolites), and hyperparathyroidism (calcium metabolism and serum PTH values).

In the case of osteosarcoma, the increase in serum alkaline phosphatase values may be useful for issuing the diagnosis. Some types of primary bone or soft tissue tumors can exceptionally cause vitamin D-resistant hypophosphatemia and osteomalacia, with regression when the tumor is removed.

Macroscopic aspects: many bone tumors, such as, for example, giant cell tumor, chondroblastoma, osteoid osteoma, chondroma and chondrosarcoma, and, for soft tissues, angioma, lipoma, etc., may have typical macroscopic features. In addition to the collection of data on the nature of the lesions, macroscopic aspects can provide further data on the aggressiveness of the tumor: infiltration and destruction of the cortex and underlying bone, level of involvement, and possible overrun, of the periosteum and soft tissues surrounding. It is very important for the pathologist to compare the macroscopic characteristics of the anatomical sample with the histological images. In this way it may be possible to understand the characteristics of the various tumor tissues, the areas of necrosis, hemorrhages, calcifications, areas of ossification, etc. This test also allows for evaluation of tumor margins and healthy tissue, reactive bone, muscle, and reactive fibrous tissue.

Biopsy: the methods of taking the bone sample require a high level of experience in clinical, diagnostic imaging, pathology and treatment of bone and soft tissue tumors. It is important to avoid the spread of cancer cells following the execution of the biopsy and also formation of hemorrhages and pathologic fractures. This can be facilitated by various factors, such as: opening of the tumor capsule; opening of the compartmental barriers (bony cortex, joint capsule, fasciae, and septa); dissection along anatomical planes; production of a hematoma that can spread along the interstitial areolar tissue, adipose tissue and muscles, even at a considerable distance from the site of the biopsy. Since all tissues can be potentially contaminated by neoplastic cells, the biopsy approach must be performed along the incision line where the next resection will be performed, or well away from the incision line that will be used for amputation. In any case, the biopsy should not open extracompartmental spaces, or intermuscular planes (use the muscle incision directly), or the areolar tissue around neuro-vascular structures, nor joint or peridural spaces. The primary decision in determining how to collect the samples is to decide whether to perform a needle aspiration, a trocar biopsy, an incisional biopsy, etc.

Needle aspiration: it is the least traumatic method and with less risk of contamination of neighboring tissues. With it we obtain: dispersion of cells or small cell groups, which can make it difficult to reach a definitive diagnosis of certainty. It is an advisable method for soft tissue tumors, which do not contain trabecular bone components and has the highest percentage of false negatives and, even when the diagnosis is positive, it can hardly be significantly accurate.

Trocar biopsy: allows obtaining samples of 3–6 cm. in diameter, which can be fixed and included, or examined in the cryostat, for histology. It is a method considered to be scarcely traumatic, which can only occasionally cause a hematoma and the spread of the tumor. The incidence of false negatives is low and it is particularly useful in those cases of tumors with accurate clinical and imaging diagnosis and where histology only needs to confirm the clinical diagnosis. This technique is recommended for all neoplastic lesions with deep localization, such as those that most frequently affect the vertebrae, the sacrum, the proximal femur, and the pelvis.

Incisional biopsy: indicated in "difficult" cases where it is necessary to evaluate detailed histological aspects and where the needle aspiration and the trocar were not sufficient to reach the final diagnosis. This technique allows for large tissue samples, but requires a great deal of experience in surgical practice.

Sections using the cryostat: the limits of this technique are those related to poor cellular detail, compared to that obtained with classic inclusion techniques. It is a practice that is generally performed in the operative phase, when it is necessary to establish whether the lesion is neoplastic or non-neoplastic, if the tumor is benign or malignant, if it has prevalent cellular characteristics referable to spindle, round, pleomorphic cells, if it is a primary tumor (e.g., osteosarcoma) or a metastasis (e.g., carcinoma).

5.4 Histology and Cytology

Bone samples to be subjected to histopathological examination should be processed with an optimal thickness of about 05–0.7 cm (Fig. 5.3).

The fixative liquid used for the fixation process, generally represented by 10% neutral buffered formalin, should always be abundant, compared to the volume of the sample itself. Some authors have also proposed the percentage of 1 volume of sample and nine volumes of fixative liquid, but, sometimes, space reasons do not allow to respect these percentages which would require the use of huge containers, especially in the case of removal of whole limbs and of large neoplastic masses. In these cases, it is advisable to take multiple samples of the neoplastic mass, to be fixed promptly and, subsequently, to send, as quickly as possible, the fixed samples, together with the "fresh" ones for anatomo-histopathological examination (Fig. 5.4).

Another important step of the diagnostic process is that of recording the samples and data, to be inserted on the container box of the material to be examined: registration number, start date of the fixation process, and start date of the subsequent decalcification process. Bone decalcification, necessary for obtaining histological sections with standard methods, involves the use of substances and/or mixtures that determine the "softening" process of the sample to be examined. This can be done with strong acids, chelating substances, and electrolysis methods. The author suggests a mixture of substances, represented by hydrochloric acid, formic acid, and distilled water, according to the proportion:

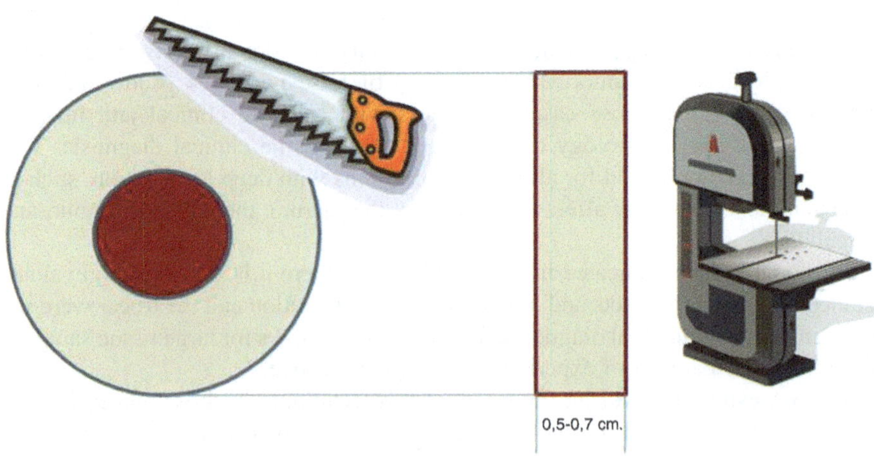

Fig. 5.3 The bandsaw for cutting bone sections

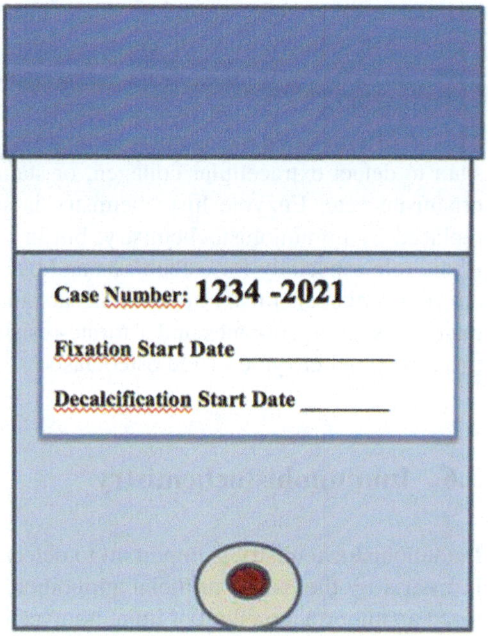

Fig. 5.4 Suitable container and label

Formic acid (99%)	Hydrochloric acid (37%)	Distilled water
50 ml	40 ml	910 ml

The descaling process must be checked daily, until the sample reaches the consistency and elasticity of the rubber and finally requires a high level of experience.

The Hematoxylin-Eosin staining, generally, is the basic technique to diagnose also a skeletal disease and may be sufficient to issue the definitive diagnosis especially when interpreted in conjunction with other clinical and instrumental data. Most bone tumors or tumor-like lesions fortunately can be diagnosed with the use of Hematoxylin-Eosin basic staining where Hematoxylin stains blue nuclei and Eosin in orange counterstain of cytoplasma and other extracellular components. Sometimes, it may be necessary to support the diagnosis with other special techniques as with immunohistochemistry.

Several basic and ancillary techniques can be used in bone pathology.

5.5 Histochemistry

Provide a variety of several stains to detect a wide range of tissutal and biochemical substances. One of the most used in bone pathology is the periodic acid-Schiff (PAS) staining useful also to detect glycogen or similar compounds in the tissues. PAS staining may be very useful in bone tumor diagnostic pathology to detect, in conjunction with the use of diastase, intracytoplasmic glycogen granules. Other useful histochemical methods including also von Kossa' staining to stain calcium, mucin staining as Alcian blue or Mayer's mucicarmine for differential diagnosis with undifferentiated metastatic adenocarcinoma, Congo Red for amyloid, trichrome stain to detect extracellular collagen, or stains to detect microorganisms infectious organisms, etc. Enzyme histochemistry is still but rarely used because has been replaced by immunohistochemistry, but in rare cases may play an important diagnostic role especially for osteoblasts and osteoclasts studies. The two main relevant enzymatic histochemistry applications are alkaline phosphatase in osteoblasts and in matrix vesicle membranes and Tartrate-resistant acid phosphatase (TRAP) to reveal this lysosomal enzyme inside osteoclasts.

5.6 Immunohistochemistry

Immunohistochemistry is important to detect extracellular and intracellular antigens in tissues by the use of artificial antibodies. This diagnostic procedure was recognized an important method of immunoinvestigations since years 1941–1942 and has emerged during the years after the production of often more highly specific antibodies and detection sophisticated systems. Immunohistochemistry is a valuable technic to identify particular tumors with "tumor markers" represented by hormones, plasma proteins, or enzymes with antigenic abilities that can be detect in tissue samples with of the bound antibody by the color of specific detection systems. Actually, the Avidin-Biotin Complex (ABC) detection system is the most used immunohistochemical method and the biotin is conjugated with secondary antibodies and work as a specific link between tissue-bound primary antibodies and the compound of avidin-biotin-peroxidase.

A similar method is the Labeled Streptavidin-Biotin (LSAB) complex that uses the secondary antibody biotin-related as a link between the streptavidin-peroxidase complex and the primary antibody (Fig. 5.5).

Usually for orthopedic investigations the specimens are cut on a cryostat or fixed in neutral buffered formalin and embedded in paraffin. For immunohistochemical studies of bone is also important to use several types of fixation to maintain good morphologic details and preserve also immunoreactivity. Alternative mixture of fixatives at neutral buffered formalin is represented by Bouin's fixative liquid or Zamboni's fixative.

Immunohistochemistry often serves to identify the histogenetic differentiation of the tumor and in particular the tissue from which they arose, but in case of anaplastic undifferentiated tumors the immunoreaction cab be bland and break down the methods.

With undifferentiated bone tumors it is necessary to resort to the application of immunohistochemical investigation protocols to establish the exact histogenesis of the neoplastic cells. Cytokeratin marker indicates epithelial origin for tumor and

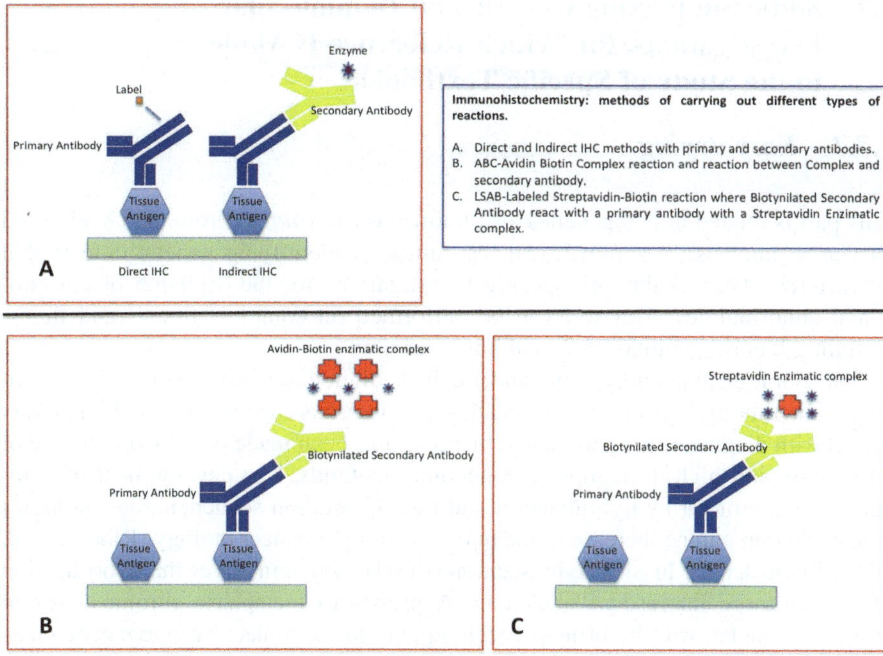

Fig. 5.5 Immunohistochemistry methods diagrams

bone metastasis while vimentin stains positive for sarcomas tumors with mesenchymal origin. Cytokeratin is very useful for differential diagnosis of metastatic carcinoma, adamantinoma, and osteofibrous dysplasia. The weak point of this technique is represented by the cases of false positives and false negatives, so it is always necessary to use specific positive and negative controls for each investigative procedure to validate the effectiveness and accuracy of the investigation. There can also be several factors that can lead to false negative or false positive results. The antibodies may not bind to the target antigens if the antigens themselves are present in the tissue, or as a result of the presence of low amounts of antigen or changes in the normal conformation of the same due to tissue processing errors, or the use of insufficient amounts of antibodies, or nonspecifically cross-reactions with other antigens, or and finally for problems related to the efficiency of the detection system. It is always suggested to perform simultaneously staining with control tissues.

Antibody's fault due to non-specific bind and cross-react with other antigens may made false positive results that in pathology may have more serious consequences than false negative.

5.7 Some Supporting Genetic and Biomolecular Investigations, for Which Reference Is Made to the Study of Specific Textbooks

5.7.1 Karyotyping

This part is related to cytogenetics also known as *Karyology* (From the Greek word *karyon* = nucleus), a sophisticated test aimed at identifying genetic defects that characterize diseases through specific investigations for the detection of chromosomal abnormalities. This test can be performed on countless tissues and fluids, including blood, amniotic fluid, and bone marrow.

The cytological phenotype of cancer cells is often described in terms of chromosomal aberrations. The rapid identification of numerous cancer potential biomarkers linked with diagnosis, prognosis, and therapies has been made possible by the use of large-scale and high-throughput cytogenomic methods. The development of comparative genomic array hybridization and next-generation sequencing technologies has resulted in a huge increase in the understanding of cancer biology (Ribeiro et al. 2019) Fluorescence In Situ Hybridization (FISH), which involves the hybridization of one or more fluorescent-labeled DNA probes to metaphase chromosomes or interphase nuclei, was the principal technique used for molecular cytogenetic characterization of chromosomes (Cui et al. 2016a). Multiplex FISH (M-FISH), spectral karyotype (SKY), and comparative genomic hybridization (CGH) are variants of the FISH technique that allow the analysis of complex and unstable tumor karyotypes (Das and Tan 2013). Nowadays, the cytogenetic analysis is based on DNA microarray technologies which involve the use of DNA probes immobilized in an array format, allowing an extremely detailed analysis of chromosomal regions without the need to have the chromosomes in metaphase as in previous technologies (Davies et al. 2005; Solinas-Toldo et al. 1997; Pollack et al. 1999). Gene expression profiling, array-CGH, single-nucleotide polymorphism array, and Next Generation Sequencing (NGS), which allow the investigation of a specific gene or the entire genome or exome, as well as gene interactions and pathways that are altered in cancer, have transformed the oncologic diagnostic field. Advances in cytogenetics and cytogenomics hold the promise of identifying precise molecular signatures connected to tumor initiation and progression, as well as viable treatment targets.

5.7.1.1 Fluorescence in Situ Hybridization (FISH)

Fluorescence In Situ Hybridization (FISH) is a cytogenetic technique used to detect and localize the presence or absence of specific sequences of DNA in chromosomes using fluorescent probes that can selectively bind a complementary sequence in specific regions of the chromosome. FISH can also be used to detect and locate specific RNA targets (mRNA, lncRNA, and miRNA) in cells, circulating tumor cells, and tissue samples. FISH is a powerful technique for mapping genes and polymorphism sites onto metaphase chromosomes and creating a physical map of the genome (Cui et al. 2016b). The FISH technique firstly involves the denaturation of the probe and the sample, the hybridization of the probe on the target cells or diffusion of the metaphase (annealing), a post-hybridization wash, and finally, the detection step, using a simple epifluorescence microscope with appropriate sets of filters (Wan 2014). FISH can be done directly on interphase nuclei, eliminating the need for time-consuming cell cultures and expanding its diagnostic applicability to include quick detection of chromosomal and genomic aberrations. An evolution of the FISH technique is represented by the multiplex FISH (M-FISH), a 24-color karyotype technique that is used for the study of complex interchromosomal rearrangements. M-FISH is based on the use of spectrally distinct fluorophores in a combinatorial way so that each homologous pair of chromosomes is uniquely labeled (Anderson 2010). Further to this, spectral karyotyping (SKY) is a multicolor-FISH technique that uses a mixture of paint probes and fluorescent dyes to determine the color emission of chromosomes. By separating a pair of different fluorescent dyes from the following five types of fluorescent dyes, Orange Spectrum, Texas Red, Cy5, Green Spectrum, and Cy5.5, and mixing them together, new colors can be created (Imataka and Arisaka 2012). The combined use of the FISH techniques and molecular biology has allowed expanding conventional cytogenetics. This type of analysis must be carried out in parallel with morphological and molecular studies in order to broaden the characterization of molecular defects in neoplasms and thus obtain an accurate diagnosis of tumors.

5.7.2 *DNA Microarray*

The basic principle behind DNA microarray is "nucleic acid hybridization." In this process, two complementary strands of DNA come together via hydrogen bonds to form a double helix molecule. This helps researchers to compare and analyze DNA or RNA molecules of identical sequences. The advancement of the DNA microarray technology has resulted in a significant amount of gene data and made it possible to track the expression patterns of thousands of genes at the same time under specific experimental settings and conditions. DNA arrays are made up of a huge number of DNA molecules that are spotted on a solid substrate in a systemic order. When the diameter of the DNA spot is less than 250 microns, DNA arrays are classified as microarrays, and when the diameter is greater than 300 microns, they are classified as macroarrays (Cho and Won 2003). Microarrays use a reverse hybridization process, which consists in fixing all the probes on support, marking the nucleic acid target to be identified, instead of marking the probe. The mRNA analysis involves a conversion to cDNA, through the use of reverse transcriptase (Brown 1999). The DNA microarray technology allows the presence of many genes within a sample to be simultaneously examined and allows the gene expression profile of a sick individual to be compared with that of a healthy one, in order to identify which genes are involved in the disease. Microarray technology represents one of the most advanced methodologies for the study of tumors as it can be applied to both the prediction and the diagnosis of cancer.

5.7.3 Polymerase Chain Reaction (PCR)

The polymerase chain reaction (PCR) is a modern and widely used molecular biology technology for reproducing DNA enzymatically without the use of living organisms, i.e. *Escherichia coli* or yeasts. Dr. Kary Mullis developed this technique in 1983; it is currently widely used and represents an essential approach in medical and biological research (Bartlett and Stirling 2003). This method allows multiple copies of DNA to be obtained starting from a DNA template. PCR can be used to diagnose a wide range of human diseases, as well as for a wide range of studies and analyses (Pui et al. 2011; Van Neste et al. 2012; Delidow et al. 1993; Genc et al. 2010). This method is now one of the most used approaches in cancer research as it allows an early diagnosis of malignant disorders such as leukemia and lymphomas. In fact, PCR is commonly employed in medical and biological research labs for a wide range of purposes, including hereditary illness detection, genetic fingerprint identification, infectious illness diagnosis, gene cloning, paternity testing, and especially to detect and diagnose a genetic issue or disease, such as cancer, by looking for specific mutations in a gene or chromosome. Hence, PCR is a significant confirmatory diagnostic tool in infectious or neoplastic diseases.

A little amount of DNA can be amplified many times in an exponential manner with this approach. The more DNA is available in the sample, the easier the analysis becomes. PCR investigations are performed using special instruments called thermal cyclers. In a thermal cycler, PCR is frequently carried out in a reaction volume of 10–200 μL in small reaction tubes. To achieve the temperatures required at each phase of the reaction, the thermal cycler cyclically heats and cools the reaction tubes.

Primer selection determines the DNA fragment to be amplified. Primers are short artificial DNA strands of less than fifty nucleotides (typically 18–25 bp) complementary to the beginning and end of the DNA fragment to be amplified. The primers binding occurs at the starting and ending points of the DNA template, where the DNA-Polymerase attaches and starts synthesis of a new DNA strand (Lung et al. 2017).

A succession of twenty to thirty-five cycles makes up the PCR process. Each one consists of three steps: (1) Denaturation (2) Annealing, and (3) Extension or Elongation.

During the denaturation step, the double-stranded DNA is heated up to 94–96 °C to break the hydrogen bonds that held the two DNA strands and thus obtains a single-stranded DNA. After the DNA strands have been separated, the temperature is decreased, from 45 to 65 °C depending on the primers, to allow the primers to connect to the single DNA strands and start the synthesis of the new strand. This process is known as annealing. Finally, the thermostable DNA-Polymerase *Taq* polymerase, so-called because it has been isolated from the thermophilic eubacterial microorganism *Thermus aquaticus*, starts the synthesis of the new DNA strands from the primers. This process, known as extension, begins with the annealed primer and progresses down the DNA strand. The temperature at which the DNA-Polymerase is extended is determined by the DNA-Polymerase used, varying approximatively from 75 to 80 °C (Delidow et al. 1993).

At the end of the PCR process, the amplicons run into an agarose gel electrophoresis where they are separated by their size. As a result, smaller DNA strands move quicker through the gel than bigger strands. The size of the PCR result can be estimated by comparing it to a DNA ladder, which is also present on the gel and contains DNA fragments of known size (Khorana et al. 1994).

5.7.4 Real-Time PCR (Quantitative PCR)

Real-Time PCR, or quantitative PCR (QRT-PCR), is a technology used to quantify nucleic acids by measuring the fluorescence emitted by a fluorophore (Bustin 2005; Mocellin et al. 2003). QRT-PCR is an indirect approach for quantifying beginning amounts of DNA, cDNA, or RNA. It is used to rapidly determine the number of PCR products, avoiding the step of agarose electrophoresis run needed in the traditional PCR method. Real-Time PCR is typically used to determine the presence of a specific DNA sequence, and the number of sequence copies in the sample. DNA is amplified by DNA-Polymerase and the formed products are monitoring continuously to quantify the DNA, thanks to sequence-specific DNA probes consisting of oligonucleotides labeled with four different dyes. The latter, only after hybridization with the DNA sample, enable the detection of the amplified DNA by the fluorescence emission (Lehmann et al. 2008). Although Real-Time PCR has high sensitivity and requires limited quantities of starting samples, the process requires a careful setting and the reagents are quite more expensive compared to those for traditional PCR. This technology has recently reached a high level of sensitivity, accuracy, and practical ease that allow this method to be used routinely for several applications. Typical uses of the QRT-PCR are the detection of pathogens, analysis of gene expression, chromosomal aberration detection, analysis of single-nucleotide polymorphism (Mai et al. 2015). Moreover, it has already been widely used in the field of cancer research as this tool can provide precious information concerning the diagnosis of several types of tumors (Mocellin et al. 2003).

DESCRIPTIVE FORM FOR THE EXAMINATION OF BONE TUMOR SPECIMENS (BIOPSY)

Case number: ______________________________

Pathologist: ______________________________ Date: ______________________________

Species: ______________________________ Breed: ______________________________

Sex: O M O F O S/N Age: __________ D/W/M/Y Weight: ____________ Lb/Kg

Site of sampling (Bone/bones involved): ______________________________

Type of biopsy sampling

______ *Excisional biopsy*

______ *Core needle biopsy*

______ *Curettage*

______ *Other (specify):* ______________________________

Number of specimens: ______________________________

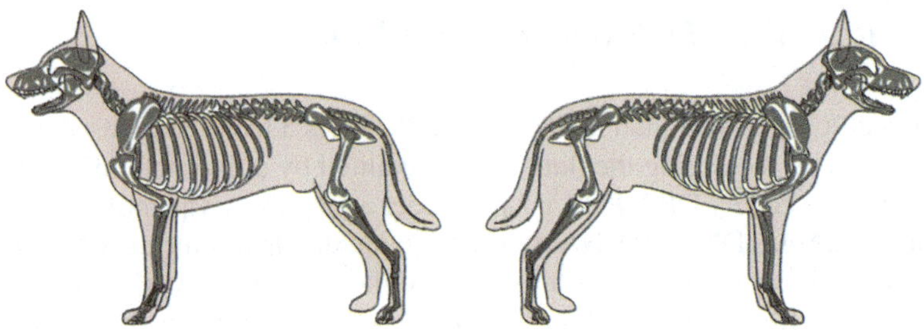

Tumor localization (select all bone lesions detected)

______ *Epiphysis*

______ *Metaphysis*

______ *Diaphysis*

______ *Cortex*

______ *Endosteum/Medullary cavity*

______ *Surface (describe also periosteal involvement, if possible)*

______ *Joint involvement*

______ *Involvement of peripheral soft tissues*

______ *No data available*

Instrumental findings: ______________________________

____

____

Gross Pathology

Shape ____________________ Texture ____________________ Color ____________________

Tumor size

Length ____________ inch/cm Width ____________ inch/cm Height ____________ inch/cm

Number of samples: ______________________________ **Number of histo-boxes**: ____________________

DESCRIPTIVE FORM FOR THE EXAMINATION OF BONE TUMOR SPECIMENS (BIOPSY)

Case number: ______________________________

Pathologist: ________________________________ Date: ____________________________

Histopathology (*WHO classification of bone tumors in domestic animals*)

Type of tumor (final diagnosis): ____

____

Mitotic index evaluated by number of mitoses/10 high-power fields (HPF): __________________

______________________ *(1HPF x 400 = 0,1734 mm2, X40 objective, evaluation of most proliferative area)*

Necrosis:

_____ *absent*

_____ *present with _____ % of extension*

_____ *impossible to evaluate*

Histologic grade (translational table for three/four grade systems to a two grade [low grade vs high grade) system

TNM Two grade system	***Three-grade system***	***Four-grade system***
Low Grade	Grade I	Grade I Grade II
High grade	Grade II Grade III	Grade III Grade IV

Invasion of vessels (vascular or lymphatic)

_____ present

_____ not identified

_____ indeterminate

Other findings: ____

Immunohistochemistry: ____________/_________________/________________/________________

____

Molecular Pathology: ____

Other: ____

____

____

____

____

____

Signature, ______________________________________ Closing date ________________________

DESCRIPTIVE FORM FOR THE EXAMINATION OF BONE TUMOR SPECIMENS (RESECTION)

Case number: ______________________________

Pathologist: ______________________________ Date: ______________________________
Species: ______________________________ Breed: ______________________________
Sex: O M O F O S/N Age: __________ D/W/M/Y Weight: __________ Lb/Kg

Site of sampling (Bone/bones involved): ______________________________

Procedure

______ *Intralesional resection*
______ *Partial/marginal resection*
______ *Wide resection*
______ *Radical resection*
______ *Other (specify):* ______________________________

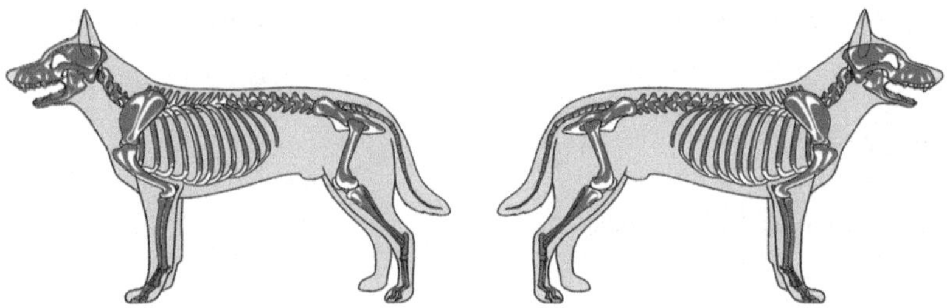

Tumor localization (select all bone lesions detected)

______ *Epiphysis*
______ *Metaphysis*
______ *Diaphysis*
______ *Cortex*
______ *Endosteum/Medullary cavity*
______ *Surface (describe also periosteal involvement, if possible)*
______ *Joint involvement*
______ *Involvement of peripheral soft tissues*
______ *No data available*

Instrumental findings: ______________________________

____

____

Gross Pathology

Shape ____________ Texture ____________ Color ____________

Tumor size

Length ________ inch/cm Width ________ inch/cm Height ________ inch/cm

Number of samples: ______________________________ **Number of histo-boxes**:______________________________

DESCRIPTIVE FORM FOR THE EXAMINATION OF BONE TUMOR SPECIMENS (RESECTION)

Case number: ______________________________

Pathologist: ______________________________ Date: ______________________

Margins:

_____ Cannot be evaluated

_____ Margins involved by the tumor (describe) ______________________________

____

_____ Margins NOT involved by the tumor

Distance of primitive mass from closet margin: ____________ cm

Descrption of Margin/s if possible: ____________________

Metastasis:

_____ not detected

_____ Vascular/Lymphatic involvement

pTNM staging:

TNM parameters

_____ m (multiple) _____ r (recurrent) _____ y (post-treatment)

Primary tumor (pT)

___ pTX not assessed ___ pT0 no evidence ___ pT1 ≤ 8cm ___ pT2 > 8cm ___ pT3 discontinous

Regional lymph nodes (pN)

___ pNX not assessed ___ pN0 no regional LN metastasis ___ PN1 regional LN metastasis

___ No LN submitted or found

Number od Lymph nodes examined ________ Specify ______________________

Distant metastasis (pM) _____ Not detetcted _____ pM1a lung _____ pM1b metastasis other lung

____

Histopathology (*WHO classification of bone tumors in domestic animals*)

Type of tumor (final diagnosis): ______________________________

____

Mitotic index evaluated by number of mitoses/10 high-power fields (HPF): ________________

________________ *(1HPF x 400 = 0,1734 mm2, X40 objective, evaluation of most proliferative area)*

Necrosis:

_____ *absent* _____ *present with* _____ *% of extension* _____ *impossible to evaluate*

Histologic grade (translational table for three/four grade systems to a two grade [low grade vs high grade) system

TNM Two grade system	*Three-grade system*	*Four-grade system*
Low Grade	Grade I	Grade I
		Grade II
High grade	Grade II	Grade III
	Grade III	Grade IV

Invasion of vessels (vascular or lymphatic)

_____ present _____ not identified _____ indeterminate

Other findings: ______________________________

Immunohistochemistry: __________/__________/__________/__________

____

Molecular Pathology: ______________________________

Other: ______________________________

____

Signature, ______________________________ Closing date ____________________

References

Anderson R (2010) Multiplex fluorescence *In Situ* hybridization (M-FISH). In: Fluorescence In Situ Hybridization (FISH). Humana Press, Totowa, NJ, pp 83–97

Bartlett JM, Stirling D (2003) A short history of the polymerase chain reaction. In: PCR protocols. Humana Press, pp 3–6

Brown PO (1999) Genome-wide analysis of DNA copy-number changes using cDNA microarrays. Nat Genet 23:41–46
Bustin SA (2005) Real-time PCR. Encyclopedia of Diagnostic Genomics and Proteomics 10(1):117
Cho SB, Won HH (2003 January). Machine learning in DNA microarray analysis for cancer classification. In: Proceedings of the First Asia-Pacific bioinformatics conference on bioinformatics, vol. 19, pp 189–198
Cui C, Shu W, Li P (2016a) Fluorescence in situ hybridization: cell-based genetic diagnostic and research applications. Front Cell Dev Biol 4:89
Cui C, Shu W, Li P (2016b) Fluorescence in situ hybridization: cell-based genetic diagnostic and research applications. Front Cell Dev Biol 4:89
Das K, Tan P (2013) Molecular cytogenetics: recent developments and applications in cancer. Clin Genet 84:315–325
Davies JJ, Wilson IM, Lam WL (2005) Array CGH technologies and their applications to cancer genomes. Chromosom Res 13:237–248
Delidow BC, Lynch JP, Peluso JJ, White BA (1993) Polymerase chain reaction. In: PCR Protocols. Humana Press, Totowa, NJ, pp 1–29
Genc A, Eroglu F, Koltas IS (2010) Detection of plasmodium vivax by nested PCR and real-time PCR. Korean J Parasitol 48(2):99
Imataka G, Arisaka O (2012) Chromosome analysis using spectral karyotyping (SKY). Cell Biochem Biophys 62(1):13–17
Khorana S, Gagel RF, Cote GJ (1994) Direct sequencing of PCR products in agarose gel slices. Nucleic Acids Res 22(16):3425
Lehmann LE, Hunfeld KP, Emrich T, Haberhausen G, Wissing H, Hoeft A, Stüber F (2008) A multiplex real-time PCR assay for rapid detection and differentiation of 25 bacterial and fungal pathogens from whole blood samples. Med Microbiol Immunol 197(3):313–324
Lung O, Ohene A, S., Buchanan, C., Joseph, T., King, R., Erickson, A., ... & Ambagala, A. (2017) Multiplex PCR and microarray for detection of swine respiratory pathogens. Transbound Emerg Dis 64(3):834–848
Mai M, Müller I, Maneg D, Lohr B, Haecker A, Haberhausen G, Hunfeld KP (2015) Real-time PCR-based identification of bacterial and fungal pathogens from blood samples. Methods Mol Biol 1237:139–147
Mocellin S, Rossi CR, Pilati P, Nitti D, Marincola FM (2003) Quantitative real-time PCR: a powerful ally in cancer research. Trends Mol Med 9(5):189–195
Pollack JR, Perou CM, Alizadeh AA, Eisen MB, Pergamenschikov A, Williams CF, Jeffrey SS, Botstein D, Brown PO (1999) Genome-wide analysis of DNA copy-number changes using cDNA microarrays. Nat Genet 23:41–46
Pui CF, Wong WC, Chai LC, Lee HY, Noorlis A, Zainazor TCT et al (2011) Multiplex PCR for the concurrent detection and differentiation of *Salmonella* spp., *Salmonella* Typhi and *Salmonella* Typhimurium. Trop Med Health 39(1):9–15
Ribeiro IP, Melo JB, Carreira IM (2019) Cytogenetics and cytogenomics evaluation in cancer. Int J Mol Sci 20(19):4711
Solinas-Toldo S, Lampel S, Stilgenbauer S, Nickolenko J, Benner A, Dohner H, Cremer T, Lichter P (1997) Matrix-based comparative genomic hybridization: biochips to screen for genomic imbalances. Genes Chromosomes Cancer 20:399–407
Van Neste L, Bigley J, Toll A, Otto G, Clark J, Delrée P, Epstein JI (2012) A tissue biopsy-based epigenetic multiplex PCR assay for prostate cancer detection. BMC Urol 12(1):1–8
Wan TS (2014) Cancer cytogenetics: methodology revisited. Ann Lab Med 34(6):413

Acknowledgments

The graphic part was handled and supported also by the architecture student of the University of Bologna-Italy, my dear son **Mr. Alberto Leonardi**.

I like to acknowledge the following friends and colleagues to support me to write this book, someone with a word of support, someone with a slide or sample, someone else with a photograph, someone with a suggestion or advice, someone with their own presence and trust, and someone with their friendship fundamental in the growth path of my life.

They are listed in alphabetical order, without other any specific order.

Thank you very much to Dr. **Giovanni Angeli** (UNIPG), Mr. **Giampaolo Ceccarani** (UNIPG), Dr. **Eleonora Calzoni** (UNIPG), Prof. Emeritus **John Edwards** (TEXAS A&M), Prof. **Fausto Elisei** (UNIPG), Prof. **Carla Emiliani** (UNIPG), Dr. **Silvia Herold** (SPRINGER), my beloved sons **Alberto and Francesco Leonardi**, Mrs. **Sara Leto** (UNIPG), Prof. **Florentina Matei** (USAMVB), Mrs. **Valeria Migni** (UNIPG), Dr. **Nicolò Montegiove** (UNIPG), Prof. **Maurizio Oliviero** (UNIPG), Mrs. **Bettye Pool, Dr. Roy Pool** (TEXAS A&M), Mr. **Bharat Sabnani** (SPRINGER), Dr. **Monica Sforna** (UNIPG), and Mr. **Luca Stefanelli** (UNIPG).

L. Leonardi, *Bone Tumors in Domestic Animals*,
https://doi.org/10.1007/978-3-030-90210-0

The manufacturer's authorised representative in the EU is Springer Nature Customer Service Centre GmbH, Europaplatz 3, 69115 Heidelberg, Germany. If you have any concerns regarding our products, please contact ProductSafety@springernature.com

Printed and bound by CPI Group (UK) Ltd, Croydon, CR0 4YY
28/09/2026
02224090-0003